AF576007

The Impact of ENT Diseases in Social Life

The Impact of ENT Diseases in Social Life

Editors

Francesco Galletti
Francesco Gazia
Francesco Freni
Cosimo Galletti

MDPI • Basel • Beijing • Wuhan • Barcelona • Belgrade • Manchester • Tokyo • Cluj • Tianjin

Editors
Francesco Galletti
Department of Adult and Development Age Human Pathology "Gaetano Barresi", unit of Otorhinolaryngology
University of Messina
Messina
Italy

Francesco Gazia
Department of Adult and Development Age Human Pathology "Gaetano Barresi", unit of Otorhinolaryngology
University of Messina
Messina
Italy

Francesco Freni
Department of Adult and Development Age Human Pathology "Gaetano Barresi", unit of Otorhinolaryngology
University of Messina
Messina
Italy

Cosimo Galletti
School of Dentistry, Campus de Bellvitge UB
University of Barcelona
Barcelona
Spain

Editorial Office
MDPI
St. Alban-Anlage 66
4052 Basel, Switzerland

This is a reprint of articles from the Special Issue published online in the open access journal *International Journal of Environmental Research and Public Health* (ISSN 1660-4601) (available at: www.mdpi.com/journal/ijerph/special_issues/ENT_diseases).

For citation purposes, cite each article independently as indicated on the article page online and as indicated below:

LastName, A.A.; LastName, B.B.; LastName, C.C. Article Title. *Journal Name* **Year**, *Volume Number*, Page Range.

ISBN 978-3-0365-1504-5 (Hbk)
ISBN 978-3-0365-1503-8 (PDF)

Contents

About the Editors

Francesco Galletti

Pr. Francesco Galletti is an ordinary professor in Otorhinolaryngology. He is the director of the UOC of Otorhinolaryngology of the G. Martino Polyclinic of the University of Messina. Expert in Otology, Cochlear Implant Surgery, infant audiology and Rhinology. Coordinator of the infantile audiological network of all of Sicily. Author of over 85 scientific articles in prestigious international high impact factor journals. His works have been cited over 500 times, and he has an H-Index of 14.

Francesco Gazia

Dr. Francesco Gazia is a specialist in Otorhinolaryngology. He works at the UOC of Otorhinolaryngology of the G. Martino Polyclinic of the University of Messina. Expert in Otology, Cochlear Implant Surgery and Rhinology. Head of the temporal bone dissection laboratory of the UOC. Author of over 23 scientific articles in prestigious international high impact factor journals. His works have been cited over 100 times, and he has an H-Index of 8.

Francesco Freni

Pr. Francesco Freni is a University researcher in Audiology. He works at the UOC of Otorhinolaryngology of the G. Martino Polyclinic of the University of Messina. Expert in Otology, Rhinology and Neck surgery. Head and Neck Cancer Committee member. Member of the melanoma committee of the Polyclinic. Author of over 60 scientific articles in prestigious international high impact factor journals. His works have been cited over 200 times, and he has an H-Index of 10.

Cosimo Galletti

Dr. Cosimo Galletti is a University researcher in Dentistry. He works at the School of Dentistry, Campus de Bellvitge UB, University of Barcelona. Expert in periodontitis, orthodontics and dental implants. Author of over 16 scientific articles in prestigious international high impact factor journals. His works have been cited over 70 times, and he has an H-Index of 5.

International Journal of
Environmental Research and Public Health

Article

Cochlear Implant Surgery: Endomeatal Approach versus Posterior Tympanotomy

Francesco Freni [1], Francesco Gazia [1,*], Victor Slavutsky [2], Enrique Perello Scherdel [3], Luis Nicenboim [4], Rodrigo Posada [5], Daniele Portelli [1], Bruno Galletti [1] and Francesco Galletti [1]

1 Department of Adult and Development Age Human Pathology "Gaetano Barresi", Unit of Otorhinolaryngology, University of Messina, 98125 Messina, Italy; franco.freni@tiscali.it (F.F.); danieleportelli@yahoo.it (D.P.); bgalletti@unime.it (B.G.); fgalletti@unime.it (F.G.)
2 Ent Clinic, 08001 Barcelona, Spain; victor.slavutsky@gmail.com
3 Servicio de ORL, Hospital Universitario Vall d'Hebron, 08001 Barcelona, Spain; 8929eps@comb.cat
4 Ear Institute, Universidad Abierta Interamericana, Rosario S2000, Argentina; entgazia@gmail.com
5 Servicio de ORL, Universidad Tecnológica de Pereira, Pereira 660001, Colombia; aagazia@gmail.com
* Correspondence: ssgazia@gmail.com; Tel.: +39-090-2212248

Received: 13 May 2020; Accepted: 9 June 2020; Published: 12 June 2020

Abstract: The aim of the present study was to compare the posterior tympanotomy (PT) technique to the endomeatal approach. The endomeatal approach (EMA) for Cochlear Implant (CI) surgery was performed on 98 patients with procident lateral sinus or a small mastoid cavity, on 103 ears (Group A). Conventional mastoidectomy and PT was performed on the other 104 patients, on 107 ears (Group B). Data on all patients were then collected for the following: intra- and post-operative complications, Tinnitus Handicap Inventory (THI), Vertigo Symptom Scale (VSS), duration of surgery, and postoperative discomfort. The difference in the total number of major and minor complications between the case group and the control group was not statistically significant. There was a statistically significant difference in discomfort between the two groups using the Visual Analogue Scale (VAS), both immediately postsurgery (p = 0.02) and after one month (p = 0.04). The mean duration of surgery was 102 ± 29 min for EMA and 118 ± 15 min for the PT technique (p = 0.008). EMA is a faster technique resulting in reduced postoperative patient discomfort in comparison to the PT method. The experience of the surgeon as well as the correct choice of surgical technique are fundamental to successful outcomes for cochlear implant surgery.

Keywords: endomeatal approach; cochlear implant; hearing loss; posterior tympanotomy; tinnitus; without mastoidectomy

1. Introduction

Cochlear Implant (CI) surgery is now the most common and reliable method used to treat patients with severe and profound sensorineural hearing loss. A CI is a surgically implanted medical device that converts acoustic sound input into electrical stimuli. The acoustic information is manipulated by the CI's speech processor to generate electrical signals, which directly stimulate the auditory nerve. Hair cells in the inner ear are not involved in the process [1–4].

The traditional surgical procedure (the transmastoid approach) involves an antromastoidectomy, followed by posterior tympanotomy (PT) through the facial recess and finally a round window technique or promontorial cochleostomy. This surgical approach has been used routinely for several years with very satisfactory results if performed by experienced otological surgeons. The main difficulty with this approach is the risk of facial nerve damage [5–8]. Alternative techniques (nonmastoidectomy approaches) have been developed in recent years. These are particularly useful when anatomic constraints are present and mean that a facial recess approach is difficult to perform. In these cases, alternative techniques must be used to minimize complications. One of the nonmastoidectomy approaches in CI surgery is the endomeatal approach (EMA). This allows an optimal and atraumatic insertion plane for the positioning of the array through the external auditory canal (EAC) and the round window (RW) [9,10].

A surgical approach using the external auditory canal and the round window as a natural access pathway for cochlear implant positioning, the endomeatal approach, is described. This approach avoids performing an antromastoidectomy, the subsequent posterior tympanotomy, and the promontorial cochleostomy. EMA requires making a bony EAC groove for electrode lead lodging in order to avoid contact between the skin and the EL that could lead to its extrusion. An overhang is left in the superior groove's edge in order to retain the electrode lead and avoid its contact with the EAC skin, therefore preventing extrusion. A 1 mm width and 2 mm depth is enough to cross the fallopian canal at a safe distance and lodge the electrode.

There are several differences between EMA and the traditional PT approach.

During the PT technique, the RW and the promontory are accessed from the back after the mastoidectomy is carried out. During the EMA procedure, the RW is accessed from a different point of view and located through the front of the posterior wall of the EAC.

A better insertion angle is obtained for EMA in comparison to PT because the scala tympani is in the same line of the insertion plane, meaning that the array does not hit the spiral lamina during introductory maneuvers. The insertion angle is approximately 30° more anterior and 15° more superior in comparison to the PT insertion angle and follows the longitudinal axis of the scala tympani. Ruptures of the modiolar wall, spiral lamina, and/or basilar membrane are avoided during EMA because the insertion line is in a more vertical position, meaning that there is enough space in the EAC to allow the array to curve over the scala tympani wall [11]. When using the traditional PT approach, there is a risk of facial nerve lesion during the mastoidectomy and PT. The incidence of this condition is around 1%. Anomalies in the facial nerve course are present in 16% of patients with CI, and one third of these have either common cavity malformations or hypoplastic cochleas with aberrant facial nerve courses [9].

The groove created during EMA enables the risk of facial nerve damage to be reduced, because the groove is in the posterior EAC wall, away from the nerve in a position easily controlled through the visualization of the Fallopian canal over the oval window.

PT through the facial recess should require facial nerve monitoring. Furthermore, the chorda tympani (CT) is often not preserved when drilling the bone, but is sometimes intentionally sacrificed in order for the surgeon to obtain a better view of the round window. During EMA, the CT is easily visualized and preserved [12].

A cholesteatoma may also develop when a PT is performed, caused by penetration of the skin in the mastoid cavity via the hole left in the posterior EAC wall. The EMA approach avoids this complication [10].

The aim of the present study was to compare the traditional transmastoid technique with EMA to determine any statistically significant differences between the two techniques for the following: intra- and postoperative complications, tinnitus, vertigo, postsurgery discomfort, and duration of intervention.

2. Materials and Methods

A retrospective multicenter study was carried out in the Department of Otorhinolaryngology at Policlinic G. Martino of Messina in Italy, the Instituto del Oído of Rosario in Argentina, and the University of Pereira in Columbia from 2005 to 2018. In total, 202 patients with deafness who had received cochlear implant surgery were enrolled, with 210 implanted ears. We included in our study all adult patients aged over 18 years with bilateral sensorineural hearing loss, who did not benefit from hearing aids, meeting at least one of these criteria:

1. A pure tone average (PTA) at 0.5, 1, 2, and 4 kHz of worse than 55 dB when tested using pure tone audiometry with bilateral hearing aids.

2. A word recognition score (WRS) of <50% recognition of bisyllabic words in open lists with optimized bilateral hearing aids.

The exclusion criteria were as follows: patients who had received previous middle ear surgery, and patients with cochlear otosclerosis or other significant craniofacial malformations. Altogether, 98 patients with procident lateral sinus, a small mastoid cavity or a narrow facial recess received the endomeatal approach (Group A), with 103 ears. An additional 104 patients with no anatomic variations, with 107 ears, received conventional mastoidectomy and TP (Group B) (Table 1). All groups received the following: intraoperative Neuronal Response Telemetry, five days of broad-spectrum antibiotics by injection, and a Stenvers projection X-ray of the skull to verify the correct positioning of the inner part of the CI the day after surgery. Activation and mapping was performed one month after surgery. We also carried out follow up assessments at 1, 3, 6, and 12 months postsurgery and once a year for the rest of life. All patients were assessed for the following: intra- and postoperative complications; Tinnitus Handicap Inventory (THI) value; Vertigo Symptom Scale (VSS) value; duration of surgery; and postoperative discomfort.

Table 1. Study population.

Features	Group A (EMA) *n* (%) M ± SD	Group B (PT) *n* (%) M ± SD	*p*-Value	Odds Ratio (95% CI)
Gender			0.50	0.82 (0.47–1.44)
Male	52/98 (53%)	60/104 (57.7%)		
Female	46/98 (47%)	44/104 (42.3%)		
Age (years)	54.35 ± 11.34	51.72 ± 8.11	0.25	
Deafness Etiology				
Genetic	50/98 (51%)	65/104 (62.5%)	0.10	0.62 (0.35–1.1)
Autoimmune	10/98 (10.2%)	13/104 (8%)	0.61	0.79 (0.33–1.9)
Infection	15/98 (15.3%)	12/104 (11.5%)	0.43	1.3 (0.61–3.1)
Idiopathic	23/98 (23.4%)	14/104 (13.4%)	0.07	1.9 (0.94–4.01)
Implant model				
Med-El	2/103 (1.9%)	3/107 (2.8%)	0.99	0.68 (0.11–4.19)
Advanced Bionics	4/103 (3.8%)	4/107 (3.7%)	0.99	1.04 (0.25–4.27)
Oticon Medical	1/103 (0.9%)	5/107 (4.6%)	0.21	0.2 (0.02–1.74)
Cochlear	96/103 (93.4%)	95/107 (88.9%)	0.33	1.73 (0.65–4.58)
PTA with aids (dB)	62.25 ± 5.30	59.84 ± 4.26	0.28	
WRS with aids (%)	60.5 ± 3.25	58.45 ± 4.65	0.35	
VSS presurgery	0.75 ± 0.28	0.88 ± 0.37	0.65	
THI presurgery	35.25 ± 10.32	38.15 ± 9.65	0.67	
Follow-up (months)	84.35 ± 15.66	82.75 ± 12.4	0.55	

n—number; %—percentage; M—media; SD—standard deviation; CI—confidence interval; PTA—Pure Tone Average; WRS—Word Recognition Score; THI—Tinnitus Handicap Inventory; VS—Vertigo Symptom Scale.

2.1. Surgical Technique

In recent times, the endomeatal approach (EMA) has been introduced for CI surgery. EMA avoids the need for mastoidectomy and PT by using the external auditory canal (EAC) and the round window (RW) as a natural access pathway for CI positioning. The surgery is carried out using the following steps:

-Incision of the skin of the EAC followed by detachment and overturn of the tympanomeatal flap.

-Prolonged retroauricular skin incision in an S-shaped cephalic direction.

-Retroauricular detachment of the skin of the posterior wall of the external third of the EAC.

-Identification of the RW, removal of the overhanging bone projection that protects the RW until complete exposure of the RW membrane.

-Formation of a bone canal 2 mm from the tympanic groove for the RW approach, proceeding towards the outer edge of the bone of the EAC between its posterior and superior wall until it reaches the temporal squama.

-An overhang is left in the superior groove's edge in order to house the electrode lead and avoid it making any contact with the skin of the EAC, thereby preventing its extrusion.

-The array is inserted into the scala tympani of the RW, following incision into the secondary tympanic membrane.

-The electrode holder filter is placed in the groove previously excavated in the EAC and stabilized with bone powder and fibrin glue.

-The body of the CI is placed in the subperiosteal pocket and secured with a titanium screw [10] (Video S1).

The traditional surgical procedure that we performed (the transmastoid approach) involves an antromastoidectomy, followed by posterior tympanotomy (PT) through the facial recess and finally a round window technique.

Both techniques have been performed by surgeons with at least 20 years of experience.

2.2. Intra- and Postoperative Complications

Intra- and postoperative complications were divided into major and minor.

Major complications included array extrusion, necrosis or severe flap infection, facial paralysis, IC failure, biofilm formation, meningitis, implant extrusion, incorrect positioning of the array, and liquorrhea from the fixing holes, requiring new intervention or causing permanent disability.

Minor complications included mild infection of the flap, hematoma, lesion of the chorda tympani with taste disturbances, selective stimulation of the facial nerve, dizziness and postoperative balance disorders, postoperative tinnitus, and perilymphatic fistula managed with drug therapy or local surgery.

2.3. Tinnitus Handicap Inventory

The Tinnitus Handicap Inventory (THI) is a self-administered test used to determine the degree of distress suffered by the patient with tinnitus. It consists of 25 questions divided into three subgroups: functional, emotional, and catastrophic. Eleven items are included on the functional scale, nine on the emotional scale, and five on the catastrophic scale. The THI uses a three-point scale: 0 (No), 2 (Sometimes), and 4 (Yes). The total score ranges from 0 to 100, and a higher score indicates a higher frequency of symptoms [13].

2.4. Vertigo Symptom Scale-Short Form

The Vertigo Symptom Scale-short form (VSS-sf) is a fifteen-item, self-administered assessment that measures the frequency of vertigo, dizziness, unsteadiness, and concomitant autonomic/anxiety symptoms over the past month. The VSS-sf uses a five-point Likert scale: 0 (never), 1 (a few times), 2 (several times), 3 (quite often/every week), and 4 (very often/most days). The total score ranges from 0 to 60, and a higher score indicates a higher frequency of symptoms [14].

2.5. Discomfort Visual Analogue Scale

The Discomfort Visual Analogue Scale (VAS) is a self-administered assessment that measures the discomfort of patients postsurgery (including pain, headache, constriction, etc.). The total score ranges from 0 to 10, and a higher score indicates a higher frequency of symptoms.

2.6. Ethical Approval

All procedures performed in studies involving human participants were in accordance with the ethical standards of the institutional and/or national research committee and with the 1964 Helsinki declaration and its later amendments or comparable ethical standards. The study was approved by the Ethics Committee of the University of Messina, on 08/27/2019 with the protocol number 81/19.

2.7. Statistical Analysis

Statistical analyses were performed using SPSS 25.0 (IBM SPSS Statistics, New York, NY, USA). The data are presented as means with standard deviations. Data normality was assessed using the Kolmogorov–Smirnov test of normality. The Mann–Whitney U-Test was used to compare the THI, VSS-sf, and Discomfort VAS measurements between the groups. Student's *t*-test was used to compare the PTA, WRS, months of follow-up, and ages of the groups. An ANOVA test with a post-hoc Tukey's HSD Test was used to compare the THI, VSS-sf, and Discomfort VAS follow-up values of each group. The chi-square test was used to compare the gender distribution of the groups as well as the etiology of deafness. The Fisher exact test was used to determine the percentage of complications that arose in each group and the implant model. Odds ratios and their corresponding 95 % Confidence Intervals (CIs) were calculated. A $p \leq 0.05$ was considered to be significant.

3. Results

Following the application of inclusion and exclusion criteria, 98 patients, of whom 52 were male and 46 were female, with a mean age of 54.35 ± 11.34, were selected for the A group. A total of 104 patients, of which 60 were male and 44 were female, with a mean age of 51.72 ± 8.11 were selected for the B group. There were no statistically significant differences between the two groups for the following: age, gender, etiology of deafness, model of implants, PTA and WRS values with aids, THI, and VSS-sf values presurgery and at follow-up.

There was only one major complication in group A, an ear infection with a biofilm formation that required CI removal, and no statistically significant difference between the two groups for this variable ($p = 0.45$, Odds Ratio = 3.1, 95% CI = 0.1–78.1). For minor complications, there were three cases of tympanic perforation in group A, and cases of subcutaneous emphysema, partial migration of the implant body, and facial nerve stimulation in group B. There was no statistically significant difference between the two groups for minor complications. The total number of minor complications was also not found to be statistically significant when comparing the case group to the control group ($p = 1$, Odds Ratio = 1.1, 95% CI = 0.2–5.2) (Table 2).

Table 2. Intra- and postoperative complications.

Complications	Group A (EMA) *n* (%)	Group B (PT) *n* (%)	*p*-Value	Odds Ratio (95% CI)
Major Complications	1 (0.9%)	0	0.45	3.1 (0.1–78.1)
Bacteria Biofilm	1 (0.9%)	0	0.45	3.1 (0.1–78.1)
Others	0	0		
Minor Complications	3 (2.9%)	3 (2.8%)	1	1.1 (0.2–5.2)
Tympanic perforation	3 (2.9%)	0	0.11	7.4 (0.3–107.7)
Subcutaneous emphysema	0	1 (0.9%)	1	0.3 (0.01–8.5)
Partial migration of the implant body	0	1 (0.9%)	1	0.3 (0.01–8.5)
Facial nerve stimulation	0	1 (0.9%)	1	0.3 (0.01–8.5)
Others	0	0		

n—number; %—percentage; CI—confidence interval

THI, VSS-sf, and Discomfort VAS values were calculated postsurgery and after 1, 3, 6, and 12 months (Table 3). There was an improvement in THI postsurgery in comparison to presurgery values ($p = 0.007$), with a further reduction in handicap after each follow-up in each of the groups ($p < 0.001$). There was no statistically significant difference between the two groups for THI scores.

There was an increase in vertiginous sensation immediately after the operation recorded on the VSS-sf, in comparison to the basal condition ($p < 0.001$). During follow-up, there was a clear improvement in these symptoms ($p < 0.001$). There was no statistically significant difference between the two groups for VSS-sf scores.

A statistically significant difference was found between the two groups for Discomfort VAS values postsurgery ($p = 0.02$) and after 1 month ($p = 0.04$). Subsequent follow-ups did not show significant differences. The mean duration of surgery was 102 ± 29 min for EMA and 118 ± 15 min for the traditional technique, with a statistically significant difference between them ($p = 0.008$).

Table 3. THI, VSS-sf, and VAS discomfort postsurgery and after 1, 3, 6, and 12-month follow-up.

Follow-Up	Group A (EMA) M ± DS	Group B (PT) M ± DS	*p*-Value
THI			
After surgery	11.8 ± 7.2	13.3 ± 8.1	0.45
1 month	9.6 ± 5.4	8.9 ± 4.6	0.55
3 months	5.4 ± 2.3	6.6 ± 2.1	0.59
6 months	2.9 ± 0.9	3.4 ± 1.2	0.21
12 months	2.1 ± 0.7	1.6 ± 0.3	0.23
VSS-sf			
After surgery	21.8 ± 9.4	19.7 ± 8.3	0.22
1 month	4.3 ± 1.2	5.8 ± 1.7	0.19
3 months	1.2 ± 0.2	1.7 ± 0.3	0.65
6 months	0.1 ± 0.02	0.1 ± 0.02	0.93
12 months	0	0	1
Discomfort VAS			
After surgery	3.2 ± 1.1	5.3 ± 1.3	**0.02**
1 month	1.9 ± 0.4	3.2 ± 0.8	**0.04**
3 months	0.2 ± 0.05	0.5 ± 0.06	0.12
6 months	0.01 ± 0.002	0.01 ± 0.002	0.95
12 months	0	0	1

M—media; SD—standard deviation; THI—Tinnitus Handicap Inventory; VSS-sf—Vertigo Symptom Scale short form; VAS—Visual Analogue Scale.

4. Discussion

The endomeatal approach (EMA) is a nonmastoidectomy approach in which cochlear implant positioning is carried out through the external auditory canal and the round window.

Our study demonstrates that no statistically significant differences were present in intra- and postoperative complications, or in the presence of tinnitus and vertigo postsurgery between groups A and B. A higher discomfort was recorded in Group B after the surgery, and a month later, whereas no statistically significant differences were present at 3, 6, and 12 months follow-up. Hospitalization time does not change between the two techniques. On average, a patient remains hospitalized for 3–4 days. This shows that both techniques are valid. The surgeon's experience with carrying out both techniques is fundamental for the success of the intervention. It is therefore advisable to evaluate on a case-by-case basis as to which of the two techniques might yield the best result.

A recent review of the literature suggests that anatomical variability is an important factor in cochlear implant surgery and is usually present [9]. The anatomy of the ear is very complex and an in-depth knowledge of this is required before carrying out surgery. The traditional approach with PT becomes more difficult in patients with procident lateral sinuses and small mastoid cavities, both of which are diagnosed with some frequency, as the surgical space for insertion of the array is reduced. With the EMA technique, insertion is safe and easy [15].

Craniofacial malformations are often present in syndromic children with congenital deafness, making traditional CI surgery with mastoidectomy and PT difficult. In the literature, EMA is also used for children with minor anatomical variations. As the child grows, there is the potential risk of dislocation of the position of the implant. The groove may then be transformed into a small furrow, which allows the passage of the guide electrode to be housed in a small mastoid fossa of about 2 cm wide, by 2 mm deep, without opening the antromastoid, and located immediately in front of the array housing. This small pit allows placement of the excess cable of the implant. Further studies on the use of EMA in children are necessary [16,17].

EMA is a less invasive technique that facilitates cochlear implant surgery in cases of anatomic differences, such as procident lateral sinus, small mastoid cavity, narrow facial recess, and an anteriorly located facial nerve.

For Tarabichi et al., there was significant variability in the relationship between the ear canal and the basal turn of the cochlea in reference to the sagittal plane. A clear majority of images demonstrated the basal turn of the cochlea to align with a more posterior angle than that of the ear canal. The trajectory provided by posterior tympanotomy aligns more favorably with the basal turn of the cochlea than transcanal access. Endoscopic technique, primarily an ear canal intervention, may not be useful in cochlear implant surgery [18].

Zernotti et al. performed a multicenter review of 208 patients with cochlear implants, comparing the different techniques. The complications were classified into major and minor. Among the 208 implanted patients, 10.5% (22 of 208) had complications. Of these, 2.88% (6 of 208) were major complications and 7.69% (16 of 208) were minor complications. Comparing the results obtained by the different approaches, the PT technique had the lowest rate of major complications (1.1%), followed by the EMA technique with 2.38% and suprameatal approach with 3.75%. As for minor complications, operations in the suprameatal approach group had the lowest rate (6.25%), followed by the EMA group (7.14%), and the group operated on using the PT technique presented the highest (10%) [19].

Mostafa et al. performed a prospective study on 125 cochlear implant patients and followed up for 6–30 months. A modified transcanal technique was adopted through a small postauricular incision. A tympanomeatal flap is elevated, the middle ear is exposed, and the round window membrane is exposed by drilling the overhanging niche. The electrode is channeled in an open trough along the posterosuperior meatal wall, which is reconstructed by autologous cartilage. The round window was used for insertion in 110 patients and a cochleostomy in 15. There were 115 complete insertions and 10 partials. There were six chorda tympani injuries, two electrode exposures with one requiring revision, and two cases with a tympanic membrane perforation which were grafted uneventfully. One case had severe infection and subsequent extrusion of the device one year after successful implantation [4].

Thaiba et al. reported the transmeatal approach that involves creating an open transcanal tunnel starting from the annulus superior to the chorda tympani laterally towards the suprameatal region. Then, through the open tunnel, a bony groove is created in the bone underneath the length of the EAC to protect the electrode from extrusion through the EAC. They described the use of this approach in 131 patients. During 2 to 46 months of follow-up, there was no electrode extrusion [20].

A shorter duration of surgery was obtained for EMA in comparison to the traditional technique, as observed by Slavutsky et al. [10,21].

A limitation of our study is that it is retrospective, with the choice of surgical technique being based on anatomical radiodiagnostic study. The study was conducted over a long period of time and in three different countries. Many surgeons were involved in the EMA and PT technique, and the medical care systems were different in each country, increasing the risk of bias. Given the limitations, our manuscript should be considered as a preliminary study. For significant results, prospective randomized double or triple-blind studies are required. Cochlear implant surgery is very delicate, regardless of the surgical technique chosen, meaning that the surgeon must have strong expertise and have mastered both techniques. Accurate presurgical planning, with careful analysis of radiological studies, is necessary when choosing the best surgical technique for each case, in order to obtain the best results from intervention.

5. Conclusions

Our multicenter study confirms that EMA is a safe surgical technique with excellent outcomes. There were no statistically significant differences with regard to complications. EMA, in our study, appears to be a faster technique with reduced postoperative patient discomfort in comparison to the more traditional PT technique. In each case, the experience of the surgeon and the correct choice of technique are fundamental in achieving a successful outcome for CI surgery.

Supplementary Materials: The following are available online at http://www.mdpi.com/1660-4601/17/12/4187/s1, Video S1: Endomeatal Approach.

Author Contributions: Conceptualization, F.G. (Francesco Galletti) and F.F.; Methodology, F.G. (Francesco Gazia); Software, D.P.; Validation, V.S., E.P.S., R.P. and L.N.; Formal analysis, F.G. (Francesco Gazia); Investigation, F.F.; Resources, D.P.; Data curation, F.G. (Francesco Gazia); Writing—original draft preparation, F.G. (Francesco Gazia) and D.P.; Writing—review and editing, B.G., F.F.; Visualization, F.G. (Francesco Galletti); Supervision, E.P.S. and V.S.; Project administration, L.N. and R.P. All authors have read and agreed to the published version of the manuscript.

Funding: This research received no external funding.

Conflicts of Interest: The authors declare no conflict of interest.

References

1. Ciodaro, F.; Freni, F.; Mannella, V.K.; Gazia, F.; Maceri, A.; Bruno, R.; Galletti, B.; Galletti, F. Use of 3D volume rendering based on high-resolution computed tomography temporal bone in patients with cochlear implants. *Am. J. Case Rep.* **2019**, *20*, 184–188. [CrossRef] [PubMed]
2. Jiang, Y.; Liu, X.; Yao, J.; Tian, Y.; Xia, C.; Li, Y.; Fu, Y.; Luo, Q. Measurement of cochlea to facial nerve canal with thin-section computed tomographic image. *J. Craniofac. Surg.* **2013**, *24*, 614–616. [CrossRef] [PubMed]
3. Roche, J.P.; Hansen, M.R. On the horizon: Cochlear implant technology. *Otolaryngol. Clin. N. Am.* **2015**, *48*, 1097–1116. [CrossRef] [PubMed]
4. Badr, E.M.; Walid, F.E.; Abdel, M.E.M. The modified transcanal approach for cochlear implantation: Technique and results. *Adv. Otolaryngol.* **2014**, *1*, 1–5.
5. Häusler, R. Cochlear implantation without mastoidectomy: The pericanal electrode insertion technique. *Acta Otolaryngol.* **2002**, *122*, 715–719. [CrossRef] [PubMed]
6. Ibrahim, I.; da Silva, S.D.; Segal, B.; Zeitouni, A. Effect of cochlear implant surgery on vestibular function: Meta-analysis study. *J. Otolaryngol. Head Neck Surg.* **2017**, *46*, 44. [CrossRef] [PubMed]
7. Manrique-Huarte, R.; Huarte, A.; Manrique, M.J. Surgical findings and auditory performance after cochlear implant revision surgery. *Eur. Arch. Otorhinolaryngol.* **2016**, *273*, 621–629. [CrossRef] [PubMed]
8. McAllister, K.; Linkhorn, H.; Gruber, M.; Giles, E.; Neeff, M. The effect of soft tissue infections on device performance in adult cochlear implant recipients. *Otol. Neurotol.* **2017**, *38*, 694–700. [CrossRef] [PubMed]
9. El-Anwar, M.W.; ElAassar, A.S.; Foad, Y.A. Non-mastoidectomy cochlear implant approaches: A literature review. *Int. Arch. Otorhinolaryngol.* **2016**, *20*, 180–184. [PubMed]
10. Slavutsky, V.; Nicenboim, L. Preliminary results in cochlear implant surgery without antromastoidectomy and with atraumatic electrode insertion: The endomeatal approach. *Eur. Arch. Otorhinolaryngol.* **2009**, *266*, 481–488. [CrossRef] [PubMed]
11. Bae, S.C.; Shin, Y.R.; Chun, Y.M. Cochlear implant surgery through round window approach is always possible. *Ann. Otol. Rhinol. Laryngol.* **2019**, *128*, 38S–44S. [CrossRef] [PubMed]
12. Hamamoto, M.; Murakami, G.; Kataura, A. Topographical relationships among the facial nerve, chorda tympani nerve and round window with special reference to the approach route for cochlear implant surgery. *Clin. Anat.* **2000**, *13*, 251–256. [CrossRef]
13. Newman, C.W.; Jacobson, G.P.; Spitzer, J.B. Development of the tinnitus handicap inventory. *Arch. Otolaryngol. Head Neck Surg.* **1996**, *122*, 143–148. [CrossRef] [PubMed]
14. Faag, C.; Bergenius, J.; Forsberg, C.; Langius-Eklöf, A. Symptoms experienced by patients with peripheral vestibular disorders: Evaluation of the vertigo symptom scale for clinical application. *Clin. Otolaryngol.* **2007**, *32*, 440–446. [CrossRef] [PubMed]
15. Galletti, F.; Freni, F.; Gazia, F.; Galletti, B. Endomeatal approach in cochlear implant surgery in a patient with small mastoid cavity and procident lateral sinus. *BMJ Case Rep.* **2019**, *12*, e229518. [CrossRef] [PubMed]
16. Sireci, F.; Ferrara, S.; Gargano, R.; Mucia, M.; Plescia, F.; Rizzo, S.; Salvago, P.; Martines, F. Hearing loss in Neonatal Intensive Care Units (NICUs): Follow-up surveillance. In *Neonatal Intensive Care Units*; Martines, F., Ed.; Nova Science Publishers: New York, NY, USA, 2017; pp. 1–15.

17. Gazia, F.; Abita, P.; Alberti, G.; Loteta, S.; Longo, P.; Caminiti, F.; Gano, R.G. NICU infants & SNHL: Experience of a western sicily tertiary care centre. *Acta Medica Mediterr.* **2019**, *35*, 1001–1007.
18. Tarabichi, M.; Nazhat, O.; Kassouma, J.; Najmi, M. Endoscopic cochlear implantation: Call for caution. *Laryngoscope* **2016**, *126*, 689–692. [CrossRef]
19. Zernotti, M.E.; Suarez, A.; Slavutsky, V.; Nicenboim, L.; Di Gregorio, M.F.; Soto, J.A. Comparison of complications by technique used in cochlear implants. *Acta Otorrinolaringol. Esp.* **2012**, *63*, 327–331. [CrossRef] [PubMed]
20. Taibah, K. The transmeatal approach: A new technique in cochlear and middle ear implants. *Cochlear Implants Int.* **2009**, *10*, 218–228. [CrossRef] [PubMed]
21. Cohen, N.L. Cochlear implant soft surgery: Fact or fantasy? *Otolaryngol. Head Neck Surg.* **1997**, *117*, 214–216. [CrossRef]

International Journal of *Environmental Research and Public Health*

MDPI

Article

Association between Chronic Obstructive Pulmonary Disease and Ménière's Disease: A Nested Case–Control Study Using a National Health Screening Cohort

So-Young Kim [1], Chang-Ho Lee [1], Dae-Myoung Yoo [2], Chan-Yang Min [2,3] and Hyo-Geun Choi [2,4,*]

1 CHA Bundang Medical Center, Department of Otorhinolaryngology-Head & Neck Surgery, CHA University, Seongnam 13496, Korea; sossi81@hanmail.net (S.-Y.K.); hearwell@gmail.com (C.-H.L.)
2 Hallym Data Science Laboratory, Hallym University College of Medicine, Anyang 14068, Korea; ydm1285@naver.com (D.-M.Y.); joicemin@naver.com (C.-Y.M.)
3 Graduate School of Public Health, Seoul National University, Seoul 08826, Korea
4 Department of Otorhinolaryngology-Head & Neck Surgery, Hallym University College of Medicine, Anyang 14068, Korea
* Correspondence: pupen@naver.com; Tel.: +82-31-380-3849

Abstract: This study explored the relation between Ménière's disease and chronic obstructive pulmonary disease (COPD). The ≥40-year-old population of the Korean National Health Insurance Service-Health Screening Cohort was included. In total, 7734 Ménière's disease patients and 30,936 control participants were enrolled. Control participants were matched for age, sex, income, and region of residence with Ménière's disease participants. The odds of having Ménière's disease given a history of COPD were analyzed using conditional logistic regression. Subgroup analyses were conducted according to age, sex, income, and region of residence. The odds of having Ménière's disease were found to be 1.18-fold higher with a history of COPD than with no history of COPD (95% confidence intervals (CI) = 1.06–1.32, E-value (CI) = 1.64 (1.31)). The ≥60 years old, male, low-income, and rural subgroups showed increased odds of developing Ménière's disease when a history of COPD was reported. A history of COPD was associated with an increased risk of Ménière's disease in the adult population.

Keywords: Ménière's disease; chronic obstructive pulmonary disease; risk factors; case–control studies; cohort studies

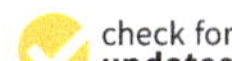

Citation: Kim, S.-Y.; Lee, C.-H.; Yoo, D.-M.; Min, C.-Y.; Choi, H.-G. Association between Chronic Obstructive Pulmonary Disease and Ménière's Disease: A Nested Case–Control Study Using a National Health Screening Cohort. *Int. J. Environ. Res. Public Health* **2021**, *18*, 4536. https://doi.org/10.3390/ijerph18094536

Academic Editor: Francesco Galletti

Received: 6 April 2021
Accepted: 21 April 2021
Published: 24 April 2021

1. Introduction

Chronic obstructive pulmonary disease (COPD) is a chronic obstructive airway disease with a prevalence of approximately 10.1% worldwide [1]. COPD is more common in men than in women, and mortality due to COPD occurs mainly in low- to middle-income countries [2]. COPD is ranked as having the third highest burden for a disease of mortality worldwide [3]. Furthermore, with an aging population, the disease burden of COPD has been increasing [4]. Although COPD is a heterogeneous and complex disease [5], progressive airway inflammation and an impaired immune response are common pathologies in patients with COPD [6,7]. COPD shares risk factors, including aging, smoking, inflammation, and physical inactivity, with other comorbidities, such as cardiovascular disease, osteoporosis, and depression [8]. In addition, COPD has been suggested to increase the risk of these comorbidities via aggravation of inflammatory and immune responses and physical inactivity [7].

Ménière's disease is described as episodic cochleovestibular symptoms of vertigo and fluctuating hearing loss [9]. Because the diagnosis of Ménière's disease depends on the symptom presentations, the early stage of Ménière's disease is difficult to identify [10]. The prevalence of Ménière's disease has been reported to be approximately 17–46 per

100,000 people [11,12]. The incidence of Ménière's disease is high in female and middle-aged populations [11,12]. Endolymphatic hydrops has been acknowledged as the primary histopathology of Ménière's disease [13]. The obstruction and/or overflow of endolymphatic flow could cause Ménière's disease [13]. The etiology of Ménière's disease has been suggested to be multifactorial, including a genetic predisposition for autoantigens and immune problems [14]. COPD has also been suggested to be caused by autoimmune problems [15]. In addition, many COPD patients suffer from neurological symptoms, such as dizziness and depression, although these symptoms are often ignored due to their chronic comorbid status. A previous study suggested the impact of chronic hypoxemia in COPD patients on cochleovestibular dysfunction [16]. Abnormalities in otoacoustic emission, hearing threshold level, and latencies of auditory brainstem response were higher in the COPD groups than in the control group [16].

The hypothesis of the present study was that COPD may be associated with the high occurrence of Ménière's disease. To test this hypothesis, we recruited patients with Ménière's disease from the national health claim database, and those with a history of COPD were compared with subjects in the control group. There has been a lack of evidence for the association of COPD with Ménière's disease. Therefore, the E-value was calculated to estimate the possible effect of unmeasured confounders [17].

2. Materials and Methods

2.1. Study Population

This study was approved by the ethics committee of Hallym University (2019-10-023). Because this study used retrospective health claims data, the requirement for written informed consent was waived by the ethics committee of Hallym University. The guidelines and regulations of the ethics committee of Hallym University were followed during all analyses. The Korean National Health Insurance Service-Health Screening Cohort data were used [18].

2.2. Participant Selection

A total of 9032 patients with Ménière's disease were selected from among 514,866 patients with 615,488,428 medical claim codes. The exclusion criteria were as follows: a diagnosis of Ménière's disease from 2002 to 2003 (n = 963); a history of treatment for head trauma (S00-S09, diagnosed by neurologists, neurosurgeons, or emergency medicine doctors) with head and neck CT evaluations (n = 275); treatment for brain tumors (C70–C72, n = 14), disorders of acoustic nerves (H933, n = 22), or benign neoplasms of cranial nerves (D333, n = 23); or patients missing total cholesterol level measurements (n = 1).

A total of 505,834 control participants who were not included in the Ménière's disease group from 2002 to 2015 were randomly selected. The exclusion criteria were as follows: patients who died before 2004; patients with no records since 2004; patients treated for Ménière's disease (ICD-10 codes: H810) without an audiometric examination (n = 16,549); patients with a history of treatment for head trauma with head and neck CT evaluations (n = 12,607); and patients treated for brain tumors (C70–C72, n = 820), acoustic nerve disorders (H933, n = 120), and benign neoplasms of cranial nerves (D333, n = 191).

The control subjects were matched 1:4 with the Ménière's disease patients for age, sex, income, and region of residence. The diagnosis date of Ménière's disease was set as the index date. Finally, 7734 Ménière's disease participants and 30,936 control participants were enrolled (Figure 1).

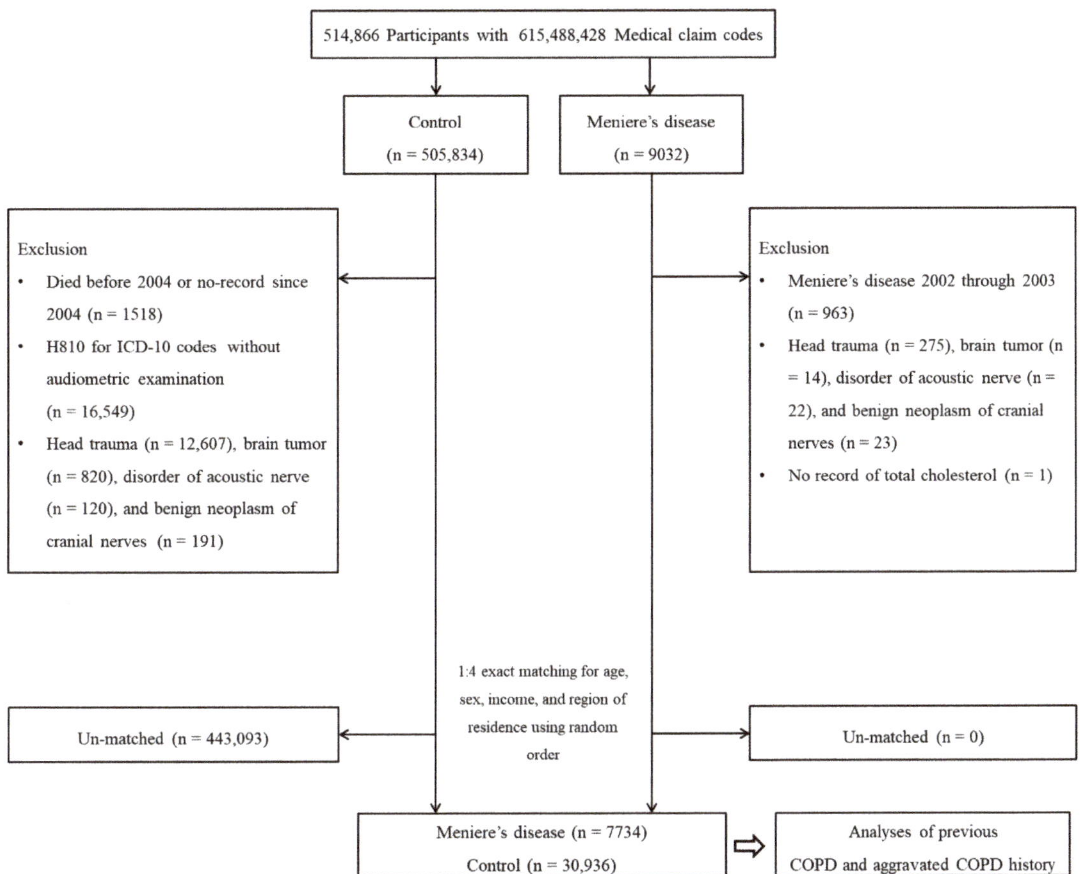

Figure 1. A schematic illustration of the participant selection process that was used in the present study. Of a total of 514,866 participants, 6054 Ménière's disease participants were matched with 24,216 control participants for age, sex, income, and region of residence. Abbreviations: COPD, chronic obstructive pulmonary disease.

2.3. Independent Variables (Chronic Obstructive Pulmonary Disease and Aggravation)

Chronic obstructive pulmonary disease was defined as the occurrence of unspecified chronic bronchitis (J42), emphysema (J43), or other COPD (J44), except MacLeod syndrome (J430) $\geq$2 times with COPD-related medications prescribed $\geq$2 times, including long-acting muscarinic antagonists (LAMAs), long-acting β2-agonists (LABAs), inhaled corticosteroids (ICSs) combined with LABAs, short-acting muscarinic antagonists (SAMAs), short-acting β2 agonists (SABAs), methylxanthine, PDE4 inhibitors, and systemic beta agonists [19].

COPD aggravation was defined as a visit to the emergency room or admission to the hospital with a COPD diagnosis.

2.4. Dependent Variable (Ménière's Disease)

Ménière's disease was included based on a history of $\geq$2 incidents of treatment for Ménière's disease (H810) and a previous audiometric examination (claim code: E6931–E6937, F6341–F6348) [20].

2.5. Covariates

The covariates included age group, income level, region of residence, obesity based on body mass index (BMI (kg/m^2)), tobacco smoking (nonsmoker, past smoker, or current smoker), and alcohol consumption (<one time a week or ≥one time a week), which were classified as described in our previous studies [21,22]. Additional measurement data, namely, total cholesterol level (mg/dL), systolic blood pressure (SBP, mmHg), diastolic blood pressure (DBP, mmHg), and fasting blood glucose level (mg/dL), were included.

Asthma was diagnosed based on a history of ≥2 incidents of treatment for asthma (J45, J46) and a history of ≥2 prescriptions for asthma medications, including ICSs or ICSs combined with LABAs, oral leukotriene antagonists (LTRAs), SABAs, systemic LABAs, xanthine derivatives, or systemic corticosteroids [21,23].

The covariates of benign paroxysmal vertigo (H811, ≥2 times), vestibular neuronitis (H812, ≥2 times), and other peripheral vertigo (H813, ≥2 times) were diagnosed based on the diagnostic codes (ICD-10). To assess the burden of comorbidities, the Charlson Comorbidity Index (CCI), except for pulmonary disease, was calculated as the continuous variable (0 (no comorbidities) to 29 (multiple comorbidities)) [24,25].

2.6. Statistical Analyses

The general characteristics were compared between the Ménière's disease and control groups using the chi-square test for categorical variables and the Wilcoxon rank-sum test for continuous variables.

Conditional logistic regression stratified by age, sex, income, and region of residence was conducted to estimate the odds ratios (ORs) and 95% confidence intervals (CIs) of COPD for Ménière's disease. The analysis was adjusted for the following covariates: asthma, obesity, smoking, alcohol consumption, CCI scores, total cholesterol level, SBP, DBP, fasting blood glucose level, benign paroxysmal vertigo, vestibular neuronitis, and other peripheral vertigo.

Subgroup analyses were conducted for age (<60 years old and ≥60 years old), sex (males and females), income (low income and high income), and region of residence (urban and rural). Additional analyses were performed in subgroups according to obesity, smoking, alcohol consumption, total cholesterol levels, blood pressure, fasting blood glucose levels, and CCI score. In these subgroups, Model 1 (adjusted for age, sex, income, and region of residence) and Model 2 (model 1 plus obesity, smoking, alcohol consumption, CCI scores, total cholesterol level, SBP, DBP, fasting blood glucose level, benign paroxysmal vertigo, vestibular neuronitis, other peripheral vertigo, and asthma) were analyzed using unconditional logistic regression.

The COPD group was divided into aggravated and non-aggravated COPD groups, and the OR of COPD for Ménière's disease was analyzed.

E-values were calculated as the sensitivity analyses [26]. SAS version 9.4 (SAS Institute Inc., Cary, NC, USA) was used for statistical analysis. All analyses were two-tailed. A p value < 0.05 was considered statistically significant.

3. Results

A total of 8.5% (660/7734) and 6.2% (1922/30,936) of Ménière's disease and control participants had histories of COPD, respectively ($p < 0.001$, Table 1). There were significant differences in the prevalence of obesity, smoking status, alcohol consumption, and CCI scores between the patients with Ménière's disease and the control group (all $p < 0.001$).

Table 1. General characteristics of participants.

Characteristics	Total Participants		
	Ménière's Disease	Control	*p*-Value
Age (years old, *n*, %)			1.000
40–44	72 (0.9)	288 (0.9)	
45–49	470 (6.1)	1880 (6.1)	
50–54	1130 (14.6)	4520 (14.6)	
55–59	1336 (17.3)	5344 (17.3)	
60–64	1257 (16.3)	5028 (16.3)	
65–69	1231 (15.9)	4924 (15.9)	
70–74	1115 (14.4)	4460 (14.4)	
75–79	726 (9.4)	2904 (9.4)	
80–84	318 (4.1)	1272 (4.1)	
85+	79 (1.0)	316 (1.0)	
Sex (n, %)			1.000
Males	2752 (35.6)	11,008 (35.6)	
Females	4982 (64.4)	19,928 (64.4)	
Income (n, %)			1.000
1 (lowest)	1343 (17.4)	5372 (17.4)	
2	967 (12.5)	3868 (12.5)	
3	1193 (15.4)	4772 (15.4)	
4	1605 (20.8)	6420 (20.8)	
5 (highest)	2626 (34.0)	10,504 (34.0)	
Region of residence (n, %)			1.000
Urban	3267 (42.2)	13,068 (42.2)	
Rural	4467 (57.8)	17,868 (57.8)	
Total cholesterol (mg/dL, mean, SD)	200.3 (38.3)	200.3 (38.4)	0.976
SBP (mmHg, mean, SD)	126.2 (16.2)	127.1 (16.9)	<0.001 †
DBP (mmHg, mean, SD)	77.6 (10.3)	78.1 (10.6)	0.001 †
Fasting blood glucose (mg/dL, mean, SD)	99.7 (25.3)	100.5 (28.2)	0.900
Obesity (n, %) ‡			<0.001 *
Underweight	152 (2.0)	823 (2.7)	
Normal	2638 (34.1)	10,997 (35.6)	
Overweight	2168 (28.0)	8305 (26.9)	
Obese I	2541 (32.9)	9803 (31.7)	
Obese II	235 (3.0)	1008 (3.3)	
Smoking status (n, %)			<0.001 *
Nonsmoker	6249 (80.8)	24,375 (78.8)	
Past smoker	813 (10.5)	3048 (9.9)	
Current smoker	672 (8.7)	3513 (11.4)	
Alcohol consumption (n, %)			<0.001 *
<1 time a week	5802 (75.0)	22,326 (72.2)	
≥1 time a week	1932 (25.0)	8610 (27.8)	
CCI score (score, n, %) §			<0.001 *
0	5088 (65.8)	21,884 (70.7)	
1	1337 (17.3)	4023 (13.0)	
2	664 (8.6)	2420 (7.8)	
3	314 (4.1)	1135 (3.7)	
≥4	331 (4.3)	1474 (4.8)	
Benign paroxysmal vertigo (n, %)	2630 (34.0)	1915 (6.2)	<0.001 *
Vestibular neuronitis (n, %)	842 (10.9)	438 (1.4)	<0.001 *
Other peripheral vertigo (n, %)	1809 (23.4)	1390 (4.5)	<0.001 *
Asthma (n, %)	1839 (23.8)	5535 (17.9)	<0.001 *
COPD (n, %)	660 (8.5)	1922 (6.2)	<0.001 *
Non-aggravated COPD	536 (6.9)	1549 (5.0)	
Aggravated COPD	124 (1.6)	373 (1.2)	

Abbreviations: CCI, Charlson comorbidity index; COPD, chronic obstructive pulmonary disease; DBP, diastolic blood pressure; SBP, systolic blood pressure; SD, standard deviation. * Chi-square test. Significance at $p < 0.05$. † Wilcoxon rank-sum test. Significance at $p < 0.05$. ‡ Obesity (BMI, body mass index, kg/m^2) was categorized as <18.5 (underweight), ≥18.5 to <23 (normal), ≥23 to <25 (overweight), ≥25 to <30 (obese I), and ≥30 (obese II). § CCI scores were calculated without pulmonary disease.

A history of COPD was associated with increased odds of having Ménière's disease (adjusted OR (aOR) = 1.18, 95% CI = 1.06–1.32, p = 0.004, E-value (CI) = 1.64 (1.31), Table 2). The positive relation of Ménière's disease with COPD was persistent in the subgroups based on age, sex, income, and region of residence (all $p < 0.05$). The aORs of a history of COPD for Ménière's disease were 1.18 (95%CI = 1.05–1.33, E-value (CI) = 1.64 (1.28)) in the ≥60 years old group, 1.28 (95%CI = 1.07–1.52, E-value (CI) = 1.88 (1.34)) in the male group, 1.26 (95%CI = 1.07–1.48, E-value (CI) = 1.83 (1.34)) in the low-income group, and 1.19 (95%CI = 1.04–1.37, E-value (CI) = 1.67 (1.24)) in the rural group.

Table 2. Odds ratios (95% confidence intervals) of chronic obstructive pulmonary disease for Ménière's disease with subgroup analyses according to age, sex, income, and region of residence.

Characteristics	No. of Ménière's Disease	Odds Ratios for Ménière's Disease				
	/No. of Participants (%)	Crude †	p-Value	Adjusted †,‡	p-Value	E-Value (CI)
Total participants (n = 38,670)						
COPD	660/2582 (25.6)	1.43 (1.30–1.57)	<0.001 *	1.18 (1.06–1.32)	0.004 *	1.64 (1.31)
Non-COPD	7074/36,088 (19.6)	1		1		
Age <60 years old (n = 15,040)						
COPD	107/381 (28.1)	1.59 (1.26–1.99)	<0.001 *	1.25 (0.95–1.63)	0.108	1.81 (1.00)
Non-COPD	2901/14,659 (19.8)	1		1		
Age ≥ 60 years old (n = 23,630)						
COPD	553/2201 (25.1)	1.40 (1.26–1.55)	<0.001 *	1.18 (1.05–1.33)	0.008 *	1.64 (1.28)
Non-COPD	4173/21,429 (19.5)	1		1		
Males (n = 13,760)						
COPD	296/1103 (26.8)	1.56 (1.35–1.80)	<0.001 *	1.28 (1.07–1.52)	0.006 *	1.88 (1.34)
Non-COPD	2456/12,657 (19.4)	1		1		
Females (n = 24,910)						
COPD	364/1479 (24.6)	1.34 (1.18–1.52)	<0.001 *	1.11 (0.96–1.28)	0.153	1.46 (1.00)
Non-COPD	4618/23,431 (19.7)	1		1		
Low income (n = 17,515)						
COPD	324/1214 (26.7)	1.53 (1.33–1.75)	<0.001 *	1.26 (1.07–1.48)	0.005 *	1.83 (1.34)
Non-COPD	3179/16,301 (19.5)	1		1		
High income (n = 21,155)						
COPD	336/1368 (24.6)	1.35 (1.18–1.53)	<0.001 *	1.11 (0.95–1.29)	0.202	1.46 (1.00)
Non-COPD	3895/19,787 (19.7)	1		1		
Urban (n = 16,335)						
COPD	229/901 (25.4)	1.41 (1.20–1.65)	<0.001 *	1.17 (0.97–1.41)	0.103	1.62 (1.00)
Non-COPD	3038/15,434 (19.7)	1		1		
Rural (n = 22,335)						
COPD	431/1681 (25.6)	1.44 (1.28–1.62)	<0.001 *	1.19 (1.04–1.37)	0.012 *	1.67 (1.24)
Non-COPD	4036/20,654 (19.5)	1		1		

Abbreviations: CCI, Charlson comorbidity index; CI, confidence interval; COPD, chronic obstructive pulmonary disease; DBP, diastolic blood pressure; SBP, systolic blood pressure. * Conditional logistic regression, significance at $p < 0.05$. † Models were stratified by age, sex, income, and region of residence. ‡ Adjusted for obesity, smoking, alcohol consumption, CCI scores, total cholesterol, SBP, DBP, fasting blood glucose, benign paroxysmal vertigo, vestibular neuronitis, other peripheral vertigo, and asthma.

Analysis of other subgroups based on obesity, smoking, alcohol consumption, total cholesterol level, blood pressure, fasting blood glucose level, and CCI score also showed an association of COPD with higher odds of having Ménière's disease (Table S1). Notably, however, there was no significant difference in the odds of having Ménière's disease between the patients with aggravated COPD and those with non-aggravated COPD (Table 3). None of the age, sex, income, or region of residence subgroups showed a definitive relation between aggravation of COPD and Ménière's disease.

Table 3. Odds ratios (95% confidence intervals) of aggravated chronic obstructive pulmonary disease for Ménière's disease with subgroup analyses according to age, sex, income, and region of residence in COPD participants.

Characteristics	No. of Ménière's Disease	Odds Ratios for Ménière's Disease				E-Value
	/No. of Participants (%)	Model 1 †	p-Value	Model 2 ‡	p-Value	Estimate (CI)
Total participants (n = 2582)						
Aggravated COPD	124/497 (25.0)	0.97 (0.76–1.22)	0.773	0.99 (0.76–1.29)	0.935	1.12 (1.00)
Non-aggravated COPD	536/2085 (25.7)	1		1		
Age <60 years old (n = 381)						
Aggravated COPD	8/31 (25.8)	0.88 (0.37–2.10)	0.774	0.75 (0.26–2.15)	0.587	2.02 (1.00)
Non-aggravated COPD	99/350 (28.3)	1		1		
Age ≥60 years old (n = 2201)						
Aggravated COPD	116/466 (24.9)	0.97 (0.76–1.24)	0.814	1.01 (0.76–1.33)	0.963	1.09 (1.00)
Non-aggravated COPD	437/1735 (25.2)	1		1		
Males (n = 1103)						
Aggravated COPD	84/304 (27.6)	1.07 (0.79–1.45)	0.655	1.19 (0.83–1.69)	0.352	1.65 (1.00)
Non-aggravated COPD	212/799 (26.5)	1		1		
Females (n = 1479)						
Aggravated COPD	40/193 (20.7)	0.82 (0.57–1.20)	0.312	0.77 (0.50–1.18)	0.226	1.93 (1.00)
Non-aggravated COPD	324/1286 (25.2)	1		1		
Low income (n = 1214)						
Aggravated COPD	59/225 (26.2)	1.04 (0.74–1.46)	0.831	0.98 (0.66–1.45)	0.907	1.18 (1.00)
Non-aggravated COPD	265/989 (26.8)	1		1		
High income (n = 1368)						
Aggravated COPD	65/272 (23.9)	0.92 (0.67–1.27)	0.604	1.04 (0.72–1.50)	0.827	1.25 (1.00)
Non-aggravated COPD	271/1096 (24.7)	1		1		
Urban (n = 901)						
Aggravated COPD	30/144 (20.8)	0.77 (0.49–1.21)	0.263	0.89 (0.53–1.49)	0.657	1.50 (1.00)
Non-aggravated COPD	199/757 (26.3)	1		1		
Rural (n = 1681)						
Aggravated COPD	94/353 (26.6)	1.05 (0.80–1.39)	0.706	1.06 (0.77–1.45)	0.737	1.30 (1.00)
Non-aggravated COPD	337/1328 (25.4)	1		1		

Abbreviations: CCI, Charlson comorbidity index; CI, confidence interval; COPD, chronic obstructive pulmonary disease; DBP, diastolic blood pressure; SBP, systolic blood pressure. Un-conditional logistic regression, significance at $p < 0.05$. † Model 1 was adjusted for by age, sex, income, and region of residence. ‡ Model 2 was adjusted for age, sex, income, region of residence, obesity, smoking, alcohol consumption, CCI scores, total cholesterol, SBP, DBP, fasting blood glucose, benign paroxysmal vertigo, vestibular neuronitis, other peripheral vertigo, and asthma.

4. Discussion

COPD was associated with an increased risk of Ménière's disease in the adult population. COPD patients showed 18% higher odds for Ménière's disease than that of the control group after adjusting for lifestyle factors of obesity, smoking, alcohol consumption, past medical histories, and laboratory findings. Other vestibular disorders of benign paroxysmal vertigo, vestibular neuronitis, and other peripheral vertigo were also adjusted. These associations were also valid in the over 60, male, low-income, and rural subgroups. The present findings proposed a novel link between COPD and Ménière's disease. Although the pathophysiology of these links between COPD and Ménière's disease has not been described, a few possible pathophysiological factors can be explained.

Autoimmunity, as a shared pathophysiology, could mediate the association of COPD with Ménière's disease. Autoimmunity has been reported to be related to Ménière's disease [27]. Approximately one-third of cases of Ménière's disease were attributed to autoimmune origin [27]. Autoimmunization to type II collagen induced endolymphatic hydrops in a guinea pig model [28]. In addition, autoantibodies against inner ear antigens were found in the sera of Ménière's disease patients [29]. Autoimmunity has also been proposed as one of the pathophysiologies of COPD [30]. The imbalance between pulmonary matrix proteases and antiproteases, which may be influenced by the generation of adaptive immune and autoimmune responses, could lead to the development of COPD [30]. The accumulation of proinflammatory T helper (Th) 1 and Th17 cells, which resemble the autoimmune response, and impaired regulatory T cell functions were observed in COPD patients [15]. In addition, autoantibodies are present in the blood of COPD patients, and autoimmunity against pulmonary antigens was observed in COPD animal models [31]. Because Ménière's disease could be diagnosed after the onset of recurrent cochleovestibular symptoms, COPD can manifest before Ménière's disease.

Ischemia may mediate the association of COPD with Ménière's disease. Insufficient inner ear blood supply has been noted as one of the pathophysiologies of inner ear dysfunction, including noise-induced hearing loss and Ménière's disease [32–34]. Because the main generator of endolymph of the cochlea involves the stria vascularis, ischemic injuries to the inner ear could impair the endolymphatic flow in the inner ear [33]. Vascular abnormalities in the lateral wall of the cochlear duct are associated with endolymphatic hydrops in guinea pigs and humans [35]. In a cadaver study, Ménière's disease patients showed occlusion of the vein of the vestibular aqueduct [36]. Notably, COPD may cause chronic ischemic status due to insufficient respiration and hypoxia. A number of previous studies demonstrated an association of COPD with myocardial infarction and stroke [37,38]. Thus, ischemia in COPD patients could hinder endolymphatic flow and induce Ménière's disease. Although the exacerbation of COPD was associated with an increased risk of myocardial infarction or stroke in previous studies, there was no association between the aggravation of COPD and Ménière's disease in the present study [37,39]. However, the relatively small size of the population with aggravated COPD (n = 124) could limit the statistical power in the present results. In addition, the high morbid state due to the aggravated COPD could deter the diagnosis of Ménière's disease.

In the subgroup analysis, COPD was associated with Ménière's disease in the elderly, male, low-income, and rural subgroups in this study. The high prevalence of COPD among elderly individuals could mediate the higher relevance of COPD to Ménière's disease [40]. In addition, the vulnerability to comorbidities due to age-related changes may influence the strong association of COPD with Ménière's disease in the elderly population. The phenotypes, symptoms, and comorbidities of COPD have been reported to show sex differences [41]. Although the incidence of COPD has shown an increasing tendency in women, men have a higher rate of COPD [41]. These sex differences could impact the association of COPD with Ménière's disease. Low socioeconomic status was also acknowledged as a risk factor for COPD [42]. Low annual income and rented housing were associated with 2.23 (95% CI = 1.97–2.53) and 1.41 (95% CI = 1.30–1.52) times higher risks of COPD, respectively [42]. The increased susceptibility to COPD may be linked with the association between COPD and Ménière's disease in these poor populations.

This study first verified the association of COPD with Ménière's disease. The large representative population provided a large control population. However, due to the health claim code database, laboratory findings, such as pulmonary function tests and vestibular function tests, were not available. The difficulty of diagnosis of Ménière's disease could have originated from the combined cochleovestibular dysfunction in as high as 71.5% of moderate to profound sensorineural hearing loss patients [43]. In addition, anatomical variation could trigger Menieriform syndrome [44]. The heterogeneous severity and types of COPD and Ménière's disease could have underestimated the association of COPD with Ménière's disease. The present study considered the lifestyle factors of smoking, obesity, and alcohol consumption. However, potential confounders, including stress level, that were not included might have influenced the current results.

5. Conclusions

A history of COPD could be a risk factor for the occurrence of Ménière's disease in the adult population. The association of COPD with Ménière's disease was solid in vulnerable populations, such as over 60, male, low-income, and rural subgroups. Clinicians should consider the co-occurrence of Ménière's disease in COPD patients who suffer from dizziness or hearing discomfort.

Supplementary Materials: The following are available online at https://www.mdpi.com/article/10.3390/ijerph18094536/s1, Table S1: Subgroup analyses of odds ratios (95% confidence intervals) of chronic obstructive pulmonary disease for Ménière's disease according to obesity, smoking, alcohol consumption, total cholesterol, blood pressure, blood glucose, and Charlson comorbidity index.

Author Contributions: Conceptualization, H.-G.C. and S.-Y.K.; methodology, H.-G.C.; formal analysis, D.-M.Y. and C.-Y.M.; investigation, D.-M.Y. and C.-Y.M.; writing—original draft preparation, H.-G.C., S.-Y.K. and C.-H.L.; writing—review and editing H.-G.C., S.-Y.K. and C.-H.L. All authors have read and agreed to the published version of the manuscript.

Funding: This work was supported in part by a research grant (NRF-2018-R1D1A1A02085328, NRF-2021-R1C1C100498611, and 2020R1A2C4002594) from the National Research Foundation (NRF) of Korea.

Institutional Review Board Statement: This study was approved by the ethics committee of Hallym University (2019-10-023). Because this study used retrospective health claims data, the requirement for written informed consent was waived by the ethics committee of Hallym University. The guidelines and regulations of the ethics committee of Hallym University were followed during all analyses.

Informed Consent Statement: Patient consent was waived due to health claim data and confirmed by the ethics committee of Hallym University.

Data Availability Statement: Release of the data by the authors is not legally allowed. All of the data are available on the database of the National Health Insurance Sharing Service (NHISS) https://nhiss.nhis.or.kr/ (accessed on 2 June 2019). NHISS permits access to all of these data via download at a certain cost for any researcher who promises to follow the research ethics.

Acknowledgments: The manuscript was edited for proper English language, grammar, punctuation, spelling, and overall style by the highly qualified native English-speaking editors at American Journal Experts (H78JFM4Q).

Conflicts of Interest: The authors declare no conflict of interest. The funders had no role in the design of the study; in the collection, analyses, or interpretation of data; in the writing of the manuscript; or in the decision to publish the results.

References

1. Celli, B.R.; Wedzicha, J.A. Update on Clinical Aspects of Chronic Obstructive Pulmonary Disease. *N. Engl. J. Med.* **2019**, *381*, 1257–1266. [CrossRef]
2. Adeloye, D.; Chua, S.; Lee, C.; Basquill, C.; Papana, A.; Theodoratou, E.; Nair, H.; Gasevic, D.; Sridhar, D.; Campbell, H.; et al. Global and regional estimates of COPD prevalence: Systematic review and meta-analysis. *J. Glob. Health* **2015**, *5*, 020415. [CrossRef]
3. Wang, H.; Naghavi, M.; Allen, C.; Barber, R.M.; Bhutta, Z.A.; Casey, D.C.; Charlson, F.J.; Chen, A.Z.; Coates, M.M.; Coggeshall, M.; et al. Global, regional, and national life expectancy, all-cause mortality, and cause-specific mortality for 249 causes of death, 1980–2015: A systematic analysis for the Global Burden of Disease Study 2015. *Lancet* **2016**, *388*, 1459–1544. [CrossRef]
4. Burney, P.G.; Patel, J.; Newson, R.; Minelli, C.; Naghavi, M. Global and regional trends in COPD mortality, 1990–2010. *Eur. Respir. J.* **2015**, *45*, 1239–1247. [CrossRef]
5. Celli, B.R.; Decramer, M.; Wedzicha, J.A.; Wilson, K.C.; Agustí, A.A.; Criner, G.J.; MacNee, W.; Make, B.J.; Rennard, S.I.; Stockley, R.A.; et al. An official American Thoracic Society/European Respiratory Society statement: Research questions in COPD. *Eur. Respir. Rev.* **2015**, *24*, 159–172. [CrossRef]
6. Barnes, P.J. Immunology of asthma and chronic obstructive pulmonary disease. *Nat. Rev. Immunol.* **2008**, *8*, 183–192. [CrossRef] [PubMed]
7. Rabe, K.F.; Watz, H. Chronic obstructive pulmonary disease. *Lancet* **2017**, *389*, 1931–1940. [CrossRef]
8. Vanfleteren, L.E.G.W.; Spruit, M.A.; Wouters, E.F.M.; Franssen, F.M.E. Management of chronic obstructive pulmonary disease beyond the lungs. *Lancet Respir. Med.* **2016**, *4*, 911–924. [CrossRef]
9. Sajjadi, H.; Paparella, M.M. Meniere's disease. *Lancet* **2008**, *372*, 406–414. [CrossRef]
10. Kim, C.-H.; Shin, J.E.; Yoo, M.H.; Park, H.J. Direction-Changing and Direction-Fixed Positional Nystagmus in Patients with Vestibular Neuritis and Meniere Disease. *Clin. Exp. Otorhinolaryngol.* **2019**, *12*, 255–260. [CrossRef]
11. Stahle, J.; Stahle, C.; Arenberg, I.K. Incidence of Meniere's disease. *Arch. Otolaryngol.* **1978**, *104*, 99–102. [CrossRef]
12. Kotimäki, J.; Sorri, M.; Aantaa, E.; Nuutinen, J. Prevalence of meniere disease in finland. *Laryngoscope* **1999**, *109*, 748–753. [CrossRef]
13. Paparella, M.M. Pathogenesis and Pathophysiology of Meniere's Disease. *Acta Oto-Laryngol.* **1991**, *111*, 26–35. [CrossRef]

14. Koyama, S.; Mitsuishi, Y.; Bibee, K.; Watanabe, I.; Terasaki, P.I. HLA Associations with Meniere's Disease. *Acta Oto-Laryngol.* **1993**, *113*, 575–578. [CrossRef]
15. Bruzzaniti, S.; Bocchino, M.; Santopaolo, M.; Calì, G.; Stanziola, A.A.; D'Amato, M.; Esposito, A.; Barra, E.; Garziano, F.; Micillo, T.; et al. An immunometabolic pathomechanism for chronic obstructive pulmonary disease. *Proc. Natl. Acad. Sci. USA* **2019**, *116*, 15625–15634. [CrossRef]
16. El-Kady, M.A.; Durrant, J.D.; Tawfik, S.; Abdel-Ghany, S.; Moussa, A.-M. Study of auditory function in patients with chronic obstructive pulmonary diseases. *Hear. Res.* **2006**, *212*, 109–116. [CrossRef] [PubMed]
17. Haneuse, S.; VanderWeele, T.J.; Arterburn, D. Using the E-Value to Assess the Potential Effect of Unmeasured Confounding in Observational Studies. *JAMA* **2019**, *321*, 602–603. [CrossRef] [PubMed]
18. Kim, S.Y.; Min, C.; Oh, D.J.; Choi, H.G. Tobacco Smoking and Alcohol Consumption Are Related to Benign Parotid Tumor: A Nested Case-Control Study Using a National Health Screening Cohort. *Clin. Exp. Otorhinolaryngol.* **2019**, *12*, 412–419. [CrossRef] [PubMed]
19. Kim, J.; Lee, J.H.; Kim, Y.; Kim, K.; Oh, Y.-M.; Yoo, K.H.; Rhee, C.K.; Yoon, H.K.; Kim, Y.S.; Park, Y.B.; et al. Association between chronic obstructive pulmonary disease and gastroesophageal reflux disease: A national cross-sectional cohort study. *BMC Pulm. Med.* **2013**, *13*, 51. [CrossRef]
20. Kim, S.Y.; Lee, C.H.; Min, C.; Park, I.; Choi, H.G. Bidirectional analysis of the association between Ménière's disease and depression: Two longitudinal follow-up studies using a national sample cohort. *Clin. Otolaryngol.* **2020**, *45*, 687–694. [CrossRef]
21. Kim, S.Y.; Min, C.; Oh, D.J.; Choi, H.G. Bidirectional Association Between GERD and Asthma: Two Longitudinal Follow-Up Studies Using a National Sample Cohort. *J. Allergy Clin. Immunol. Pract.* **2020**, *8*, 1005–1013. [CrossRef]
22. Kim, S.Y.; Oh, D.J.; Park, B.; Choi, H.G. Bell's palsy and obesity, alcohol consumption and smoking: A nested case-control study using a national health screening cohort. *Sci. Rep.* **2020**, *10*, 4248. [CrossRef]
23. Kim, S.; Kim, J.; Kim, K.; Kim, Y.; Park, Y.; Baek, S.; Park, S.Y.; Yoon, S.-Y.; Kwon, H.-S.; Cho, Y.S.; et al. Healthcare use and prescription patterns associated with adult asthma in Korea: Analysis of the NHI claims database. *Allergy* **2013**, *68*, 1435–1442. [CrossRef]
24. Quan, H.; Li, B.; Couris, C.M.; Fushimi, K.; Graham, P.; Hider, P.; Januel, J.-M.; Sundararajan, V. Updating and Validating the Charlson Comorbidity Index and Score for Risk Adjustment in Hospital Discharge Abstracts Using Data From 6 Countries. *Am. J. Epidemiol.* **2011**, *173*, 676–682. [CrossRef]
25. Quan, H.; Sundararajan, V.; Halfon, P.; Fong, A.; Burnand, B.; Luthi, J.-C.; Saunders, L.D.; Beck, C.A.; Feasby, T.E.; Ghali, W.A. Coding Algorithms for Defining Comorbidities in ICD-9-CM and ICD-10 Administrative Data. *Med. Care* **2005**, *43*, 1130–1139. [CrossRef]
26. VanderWeele, T.J.; Ding, P. Sensitivity Analysis in Observational Research: Introducing the E-Value. *Ann. Intern. Med.* **2017**, *167*, 268–274. [CrossRef] [PubMed]
27. Greco, A.; Gallo, A.; Fusconi, M.; Marinelli, C.; Macri, G.F.; De Vincentiis, M. Meniere's disease might be an autoimmune condition? *Autoimmun. Rev.* **2012**, *11*, 731–738. [CrossRef]
28. Yoo, T.; Yazawa, Y.; Tomoda, K.; Floyd, R. Type II collagen-induced autoimmune endolymphatic hydrops in guinea pig. *Science* **1983**, *222*, 65–67. [CrossRef]
29. Yoo, T.J.; Ge, X.; Sener, O.; Mora, M.; Kwon, S.S.; Mora, F.; Barbieri, M.; Shea, J.; Yazawa, Y.; Mora, R.; et al. Presence of autoantibodies in the sera of Meniere's disease. *Ann. Otol. Rhinol. Laryngol.* **2001**, *110 Pt 1*, 425–429. [CrossRef]
30. Stefanska, A.M.; Walsh, P.T. Chronic Obstructive Pulmonary Disease: Evidence for an Autoimmune Component. *Cell. Mol. Immunol.* **2009**, *6*, 81–86. [CrossRef]
31. Wen, L.; Krauss-Etschmann, S.; Petersen, F.; Yu, X. Autoantibodies in Chronic Obstructive Pulmonary Disease. *Front. Immunol.* **2018**, *9*, 66. [CrossRef]
32. Thorne, P.; Nuttall, A. Laser doppler measurements of cochlear blood flow during loud sound exposure in the guinea pig. *Hear. Res.* **1987**, *27*, 1–10. [CrossRef]
33. Yazawa, Y.; Kitano, H.; Suzuki, M.; Tanaka, H.; Kitajima, K. Studies of cochlear blood flow in guinea pigs with endolymphatic hydrops. *ORL* **1998**, *60*, 4–11. [CrossRef] [PubMed]
34. Huh, G.; Bae, Y.J.; Woo, H.J.; Park, J.H.; Koo, J.-W.; Song, J.-J. Vestibulocochlear Symptoms Caused by Vertebrobasilar Dolichoectasia. *Clin. Exp. Otorhinolaryngol.* **2020**, *13*, 123–132. [CrossRef]
35. Nakai, Y.; Masutani, H.; Moriguchi, M.; Matsunaga, K.; Kato, A.; Maeda, H. Microvasculature of normal and hydropic labyrinth. *Scanning Microsc.* **1992**, *6*, 1097.
36. Friberg, U.; Rask-Andersen, H. Vascular occlusion in the endolymphatic sac in Meniere's disease. *Ann. Otol. Rhinol. Laryngol.* **2002**, *111 Pt 1*, 237–245. [CrossRef]
37. Donaldson, G.C.; Hurst, J.R.; Smith, C.J.; Hubbard, R.B.; Wedzicha, J.A. Increased Risk of Myocardial Infarction and Stroke Following Exacerbation of COPD. *Chest* **2010**, *137*, 1091–1097. [CrossRef]
38. Corlateanu, A.; Covantev, S.; Mathioudakis, A.G.; Botnaru, V.; Cazzola, M.; Siafakas, N. Chronic Obstructive Pulmonary Disease and Stroke. *COPD J. Chronic Obstr. Pulm. Dis.* **2018**, *15*, 405–413. [CrossRef]
39. Reilev, M.; Pottegård, A.; Lykkegaard, J.; Søndergaard, J.; Ingebrigtsen, T.S.; Hallas, J. Increased risk of major adverse cardiac events following the onset of acute exacerbations of COPD. *Respirology* **2019**, *24*, 1183–1190. [CrossRef]

40. Cortopassi, F.; Gurung, P.; Pinto-Plata, V. Chronic Obstructive Pulmonary Disease in Elderly Patients. *Clin. Geriatr. Med.* **2017**, *33*, 539–552. [CrossRef]
41. Raghavan, D.; Varkey, A.; Bartter, T. Chronic obstructive pulmonary disease. *Curr. Opin. Pulm. Med.* **2017**, *23*, 117–123. [CrossRef]
42. Borné, Y.; Ashraf, W.; Zaigham, S.; Frantz, S. Socioeconomic circumstances and incidence of chronic obstructive pulmonary disease (COPD) in an urban population in Sweden. *COPD J. Chronic Obstr. Pulm. Dis.* **2019**, *16*, 51–57. [CrossRef]
43. Ciodaro, F.; Freni, F.; Alberti, G.; Forelli, M.; Gazia, F.; Bruno, R.; Sherdell, E.P.; Galletti, B.; Galletti, F. Application of Cervical Vestibular-Evoked Myogenic Potentials in Adults with Moderate to Profound Sensorineural Hearing Loss: A Preliminary Study. *Int. Arch. Otorhinolaryngol.* **2019**, *24*, e5–e10. [CrossRef]
44. Galletti, F.; Freni, F.; Gazia, F.; Galletti, B. Endomeatal approach in cochlear implant surgery in a patient with small mastoid cavity and procident lateral sinus. *BMJ Case Rep.* **2019**, *12*, e229518. [CrossRef]

International Journal of
Environmental Research and Public Health

MDPI

Article

The Use of Hyperbaric Oxygen Therapy and Corticosteroid Therapy in Acute Acoustic Trauma: 15 Years' Experience at the Czech Military Health Service

Richard Holy [1,2], Sarka Zavazalova [1,2], Klara Prochazkova [2,3], David Kalfert [4,*], Temoore Younus [1,2], Petr Dosel [5], Daniel Kovar [1,2], Karla Janouskova [1,2], Boris Oniscenko [5], Zdenek Fik [4] and Jaromir Astl [1,2]

1 Department of Otorhinolaryngology and Maxillofacial Surgery, Military University Hospital, 16902 Prague, Czech Republic; richard.holy@uvn.cz (R.H.); sarka.zavazalova@uvn.cz (S.Z.); tny1996@gmail.com (T.Y.); daniel.kovar@uvn.cz (D.K.); janouskova.karla@uvn.cz (K.J.); jaromir.astl@uvn.cz (J.A.)
2 Third Faculty of Medicine, Charles University, 10000 Prague, Czech Republic; klara.prochazkova@fnkv.cz
3 Department of Otorhinolaryngology, University Hospital Kralovské Vinohrady, 10034 Prague, Czech Republic
4 Department of Otorhinolaryngology and Head and Neck Surgery, University Hospital Motol, First Faculty of Medicine, Charles University, 15006 Prague, Czech Republic; zdenek.fik@fnmotol.cz
5 The Institute of Aviation Medicine, 16000 Prague, Czech Republic; petrdosel@atlas.cz (P.D.); oniscenko@ulz.cz (B.O.)
* Correspondence: david.kalfert@fnmotol.cz; Tel.: +420-224434311

Citation: Holy, R.; Zavazalova, S.; Prochazkova, K.; Kalfert, D.; Younus, T.; Dosel, P.; Kovar, D.; Janouskova, K.; Oniscenko, B.; Fik, Z.; et al. The Use of Hyperbaric Oxygen Therapy and Corticosteroid Therapy in Acute Acoustic Trauma: 15 Years' Experience at the Czech Military Health Service. *Int. J. Environ. Res. Public Health* **2021**, *18*, 4460. https://doi.org/10.3390/ijerph18094460

Academic Editors: Francesco Galletti, Francesco Gazia, Francesco Freni and Cosimo Galletti

Received: 24 March 2021
Accepted: 21 April 2021
Published: 22 April 2021

Abstract: Background: Acute acoustic trauma (AAT) ranks, among others, as one common cause of inner ear function impairment, especially in terms of military personnel, who are at an increased exposure to impulse noises from firearms. Aim of this study: 1. We wanted to demonstrate whether early treatment of AAT means a higher chance for the patient to improve hearing after trauma. 2. We find the answer to the question of whether hyperbaric oxygen therapy (HBO2) has a positive effect in the treatment of AAT. Methods: We retrospectively analyzed data for the period 2004–2019 in patients with AAT. We evaluated the therapeutic success of corticosteroids and HBO2 in a cohort of patients with AAT n = 108 patients/n = 141 affected ears. Results: Hearing improvement after treatment was recorded in a total of 111 ears (79%). In terms of the data analysis we were able to ascertain, utilizing success of treatment versus timing: within 24 h following the onset of therapy in 56 (40%) ears—54 (96%) ears had improved; within seven days following the onset the therapy was used in 55 (39%) ears—41 (74%) ears had improved; after seven days the therapy started in 30 (21%) ears—16 (53%) ears had improved. Parameter latency of the beginning of the treatment of AAT was statistically significant (p = 0.001 and 0.017, respectively). The success of the medical protocols was apparent in both groups—group I (treated without HBO2): n = 61 ears, of which 50 (82%) improved, group II (treated with HBO2): n = 73 ears, of which 56 (77%) improved. Group II shows improvement at most frequencies (500–2000 Hz). The most serious sensorineural hearing loss after AAT was at a frequency of 6000 Hz. Conclusion: Analysis of our data shows that there is a statistically significant higher rate of improvement if AAT treatment was initiated within the first seven days after acoustic trauma. Early treatment of AAT leads to better treatment success. HBO2 is considered a rescue therapy for the treatment of AAT. According to our recommendation, it is desirable to start corticosteroid therapy immediately after acoustic trauma. If hearing does not improve during the first seven days of corticosteroid therapy, then HBO2 treatment should be initiated.

Keywords: acute acoustic trauma; noise induced hearing loss; tinnitus; hyperbaric oxygen therapy

1. Introduction

During acute acoustic trauma (AAT), the inner ear becomes mechanically damaged, after a short-impact acoustic impulse (intensity of 90–130 dB for a duration of 1 ms).

In terms of pathology, protective middle ear reflexes are blocked, which cause an alteration of action potential formation. Vasospasm of microcirculation and hypoxia of sensory cells occur, in order to prevent metabolic imbalance. Pathological processes may result in damaging hair cells and dendrites of primary auditory neurons that consequently induce a transition stage between the regeneration and cell death. This so-called transition stage may primarily influence the line of therapy [1].

Typical AAT symptoms include high-frequency sensorineural hearing loss (4 kHz and higher, while 1–2 kHz influenced minimally) and tinnitus.

Vertigo or spontaneous nystagmus are rarely present [2].

There are different severities of AAT, for instance, less severe onset (which is less frequent) causes reversible hearing loss—temporary threshold shifts (TTS)—hearing is restored within 24 h from acoustic trauma [3,4]. More severe onset (which is more frequent) causes irreversible hearing loss—permanent threshold shifts (PTS) [3–5].

The optimal treatment for AAT is yet to be defined [3–5]. Some animal studies display that hyperbaric oxygen therapy (HBO2) combined with corticosteroid therapy improve the functional and morphological conditions of the inner ear, by allowing permanent therapeutic effect through noise-induced cochlear hypoxia [3–5]. However, the negative effect of HBO2 is also reported in the literature (animal study) [6]. Some papers states that expectant non-interventional recovery of hearing does not belong among therapeutic alternatives as in most cases hearing recovery is incomplete. After AAT a partial hearing loss and tinnitus usually persist. On the other hand, it is described in literature—spontaneous hearing recovery without the treatment. It can probably be attributed to a naturally occurring phenomena [4,7]. Kuznecov et al. present in their manuscript the groups of the most used drugs for the pharmacological correction of hearing loss after AAT: corticosteroids, antioxidants, nootropics, antihypoxants, and others [8].

Administration of therapy (of AAT) is quintessentially begun within 24 h after the acoustic trauma [9–11]. Using the data analysis of our group of patients undergoing standardized medical protocols (corticosteroid therapy, HBO2 therapy), we aim to prove a causal connection between starting early treatment and a better prognosis for hearing loss improvement, thus proving our objective [9–11].

Although 75% of AAT of cases can be classified as occupational accidents, an absolutely necessary precondition is good awareness and cooperation of military general practitioners, through immediate administration of corticosteroid therapy at the site of trauma with prompt referral of soldiers to the relevant military hospital for further therapy (corticosteroid therapy, vasodilator treatment, and HBO2 therapy). For instance, in Finland, just as in the Czech Republic, several hundred soldiers suffer from AAT every year despite strict security regulations dealing with shooting from firearms in defense forces, in the USA, 20–30% of soldiers experience hearing impairment [1,9,12–15].

Issues of HBO2 Therapy in AAT from the Perspective of a Hyperbaric Medicine Expert

Hyperbaric oxygen therapy is a type of inhalation treatment using highly concentrated oxygen inside a hyperbaric chamber, in which the pressure is higher than atmospheric pressure. The therapeutic pressures range between 200–280 kPa (2–2.8 ATA = absolute technical atmosphere). The therapeutic excess pressures range between 100–180 kPa (1–1.8 ATA). The usual treatment exposure time is 120 min. Patients with hearing loss are exposed to HBO2 therapy once a day [8,9,16].

HBO2 therapy contributes to AAT treatment by improving oxygenation of the inner ear, which results in the adjustment of transmembrane potential, activation of cell metabolism, and regeneration of ionic balance. Rheologically, the effect of oxygen diffusing through the oval window leads to a decrease in hematocrit and blood viscosity [8–10].

Several ear disorders correlated in literature with anaerobic bacteria infection, influencing the prognosis of serious diseases such as lateral cervical and mediastinal involvements; however excellent response to hyperbaric therapy combined with antibiotics and cortisone drugs has been reported [17,18].

Aim of this study: Demonstrate whether early treatment of AAT means a higher chance of improving hearing after trauma. Answer the question of the positive effect of HBO2 in the treatment of AAT.

2. Materials and Methods

We retrospectively analyzed data from n = 108 patients/n = 141 damaged ears (33 patients with bilateral AAT) treated in the period between 2004 and 2019. This cohort of patients consisted of 97 men (90%) and 11 women (10%). A total of 65 representatives of group A (soldiers) (60%), 43 representatives of group B (civilians) (40%). The age range was between 18 and 82 years, the average age was 38 years.

The etiology of the AAT source was recorded: after shooting 102 (72%), after an explosion 11 (8%), after a music concert 10 (7%), and others 18 (13%) (after a barking dog, after whistling, after shouting, after car battery explosion, after incorrect fitting of earplugs/earmuffs).

2.1. Division into Groups According

We divided the cohort of patients into groups:

Group A—were soldiers (sound intensity at AAT was up to 170 dB)
Group B—were civilian persons (sound intensity at AAT was up to 120 dB)

We also divided the cohort of patients into groups according to the start of treatment:

Parameter latency of the beginning of the treatment of AAT within 24 h
Parameter latency of the beginning of the treatment of AAT within 7 days
Parameter latency of the beginning of the treatment of AAT after 7 days

Further division of the patient cohort into groups was according to the method of treatment:

Group I—patients were treated with corticosteroids + vasodilatory infusion, without hyperbaric oxygen therapy. In this group, the age range was between 20 and 82 years, the average age was 33 years.
Group II—patients were treated with corticosteroids + vasodilatory infusion + hyperbaric oxygen therapy. In this group, The age range was between 18 and 69 years, the average age was 38 years.

Whilst monitoring the effect of particular medical protocols we singled out a group of patients ("Singled out group p. o. vasodilatants"— who were treated only by vasodilators p.o. (betahistin-dihydrochlorid 24 mg, vinpocetin), i.e., 7 damaged ears. In comparison with the effect of the medical protocol without HBO2 vs. with HBO2, we finally evaluated a group of 134 damaged ears. Audiometric measurements were taken with the help of pure tone audiometry (PTA) (measurement dB HL). We used the device Orbiter 992 with Headphone TDH 39 and with the valid Certificate from the Metrology Department, Czech Republic. Measurements were taken before and at the end of treatment. First PTA was performed by soldiers within 24 h after AAT. First PTA was performed by civilians as soon as they patients visit our clinic. Last PTA was performed two months after finishing therapy.

2.2. Medical Protocol

On the day of AAT or on the day of the first examination—corticosteroid therapy started with Solu-Medrol (1st day: Solu-Medrol 125 mg; 2nd day: Solu-Medrol 80 mg; 3rd day: Solu-Medrol 40 mg; in 100 mL of physiological solution i.v.) in addition to vasodilation infusions (20 mg of ethyl apovincaminate alias known as "vinpocetine"; in 250 mL of physiological solution; 120 min; 10 days). In the case there was no amelioration of the condition within 7 days (verified by PTA), we decided to start hyperbaric oxygen therapy as soon as possible (10 exposures in the hyperbaric chamber in the Institute of Aviation Medicine, Prague; pressure 2.5 atmosphere for 120 min). A special group of patients (n = 7 ears) were treated only with vasodilators peroral—vinpocetine 10 mg one tablet twice per day, betahistin-dihydrochlorid 24 mg one tablet three times per day.

The standard medical protocol of HBO2 therapy used in this study includes a compression phase during which the pressure in the hyperbaric chamber is increased for a period of 15 min to the therapeutic level of 250 kPa (2.5 ATA). Both initial compression phase, from surface (1.0 ATA) to the treatment depth (250 kPa = 2.5 ATA), and the final decompression from depth to surface are in, they last 30 min together (15 min/each at a compression/deco-speed of 1 m/min). The complete 'dive' table used in this study took 90 min, including an interposed break of 5 min in ambient air at depth (such a break is applied as conservative prevention from hyperbaric oxygen possible side effects). During compression, the patient is ventilating atmospheric air (within the chamber), because wearing an oxygen mask would obstruct performing active maneuvers for equalizing the increase of pressure to the middle ear. Compression relates to the most frequent occurrence of baric problems (usually earache or pain of paranasal sinuses) because of insufficient ventilation function of the Eustachian tube. Final decompression lasts 15 min and there are usually no baric problems. The standard minimum number of Hyperbaric Oxygen Treatments (Tx) required in this case is 10 [2,8,9].

2.3. Statistical Analysis of Data

All statistical analyses were performed with IBM SPSS Statistics (version 22.0; SPSS, IBM, Armonk, NY, USA). We used the non-parametric Mann–Whitney test and Fisher's exact test and p value < 0.05 was used to establish statistical significance.

3. Results

Results of lateral prevalence of AAT did not show a greater difference in significance between Group A and Group B. Bilateral damage occurred in 33 patients (30%).

Hearing improved in statistically younger patients (average age 34, 4 years; level of significance $p = 0.001$) as compared to older ones (average 44, 4 years).

Table 1 shows the success rate of AAT treatment. In total, 79% improved, Group A vs. Group B, 70% vs. 81%, respectively. Hearing improvement in standard terms (when hearing loss threshold after treatment was above 20 dB HL) was 41% in total, Group A vs. Group B, 39% vs. 29%, respectively. In group B, there was a statistically significant (at $p = 0.012$) higher partial hearing improvement after AAT treatment.

Table 1. Success of AAT treatment, All AAT, Group A (soldiers), Group B (civilians person).

	All AAT		**Group A**		**Group B**		***p*-Value ***
Total Number of Damaged Ears	**n = 141**		**n = 83**		**n = 58**		
Improved in total	111	79%	58	70%	47	81%	0.096
Restored to standard = after treatment normacusis	58	41%	32	39%	17	29%	0.170
Partially improved	53	38%	26	31%	30	52%	0.012
Not improved	30	21%	25	30%	11	19%	

* Fisher's exact test.

Within the group of patients with improved conditions, the figures of sound intensity before the treatment, in all frequencies, were significantly lower than in non-improved, on the level of significance 0.05 resp. 0.1 (frequency 1000, 4000, 6000 Hz).

An important parameter, early treatment of AAT treatment, is shown in Table 2. Improvement of hearing—after the treatment started within 24 h was 96%, Group A 97% and Group B 95%. After the treatment started within seven days, improvement of hearing was in 74%, 79% at group A and 70% at group B. More than seven days after treatment started, improvement of hearing was in 53%, 53% in Group A and 55% in Group B. Parameter: Latency of the beginning of the treatment—whole set, Group A and Group B, this parameter of early treatment of AAT is always statistically significant ($p \leq 0.001$ and 0.017, respectively).

Table 2. (**a**) Latency of the beginning of the treatment—the whole set (total AAT). (**b**) Latency of the beginning of the treatment—Group A (soldiers). (**c**) Latency of the beginning of the treatment—Group B (civilian persons).

(a)			
Latency of the Beginning of the Treatment	**Total AAT n = 141 Ears**	**Improved after Treatment**	***p*-Value**
Within 24 h	n = 56; 40%	n = 54; 96%	
Within 7 days	n = 55; 39%	n = 41; 74%	<0.001
After 7 days	n = 30; 21%	n = 16; 53%	
(b)			
Latency of the Beginning of the Treatment	**Group A n = 83 Ears**	**Improved after Treatment**	***p*-Value**
Within 24 h	n = 36; 43%	n = 35; 97%	<0.001
Within 7 days	n = 28; 34%	n = 22; 79%	
After 7 days	n = 19; 23%	n = 10; 53%	
(c)			
Latency of the Beginning of the Treatment	**Group B n = 58 Ears**	**Improved after Treatment**	***p*-Value**
Within 24 h	n = 20; 34%	n = 19; 95%	0.017
Within 7 days	n = 27; 47%	n = 19; 70%	
After 7 days	n = 11; 19%	n = 6; 55%	

* Fisher's exact probability test.

The earlier AAT treatment was started, the higher were the chances of hearing improvement ($p \leq 0.001$ and 0.017, respectively)

Figures 1–3 show the figures of PTA curve before and after the treatment in total, in Group A and Group B. Total improvement of the hearing threshold after treatment, at the highest affected frequencies, was by 8 dB. The charts show clearly that the greatest damage of the whole set, both group A and group B, was at 6000 Hz and in descending order at 8000, 4000, 2000 Hz, respectively.

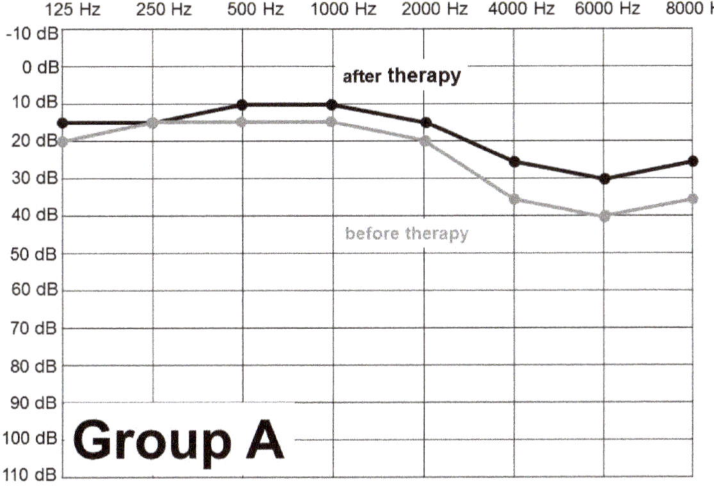

Figure 1. PTA before and after the treatment of AAT—Group A—soldiers (n = 83).

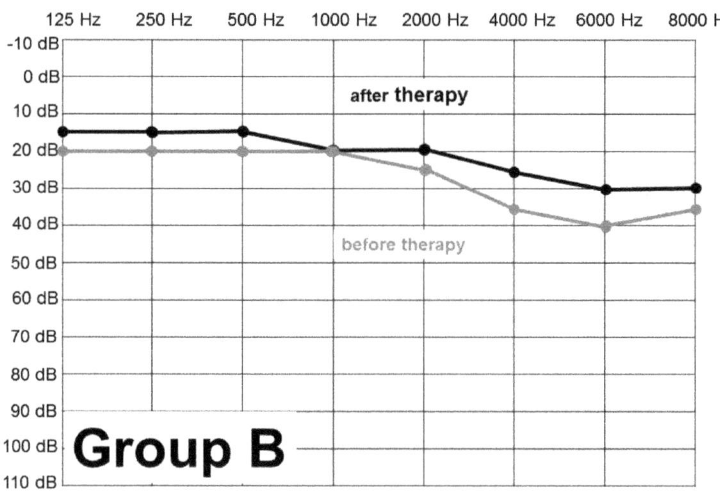

Figure 2. PTA before and after the treatment of AAT—Group B—civilian persons (n = 58).

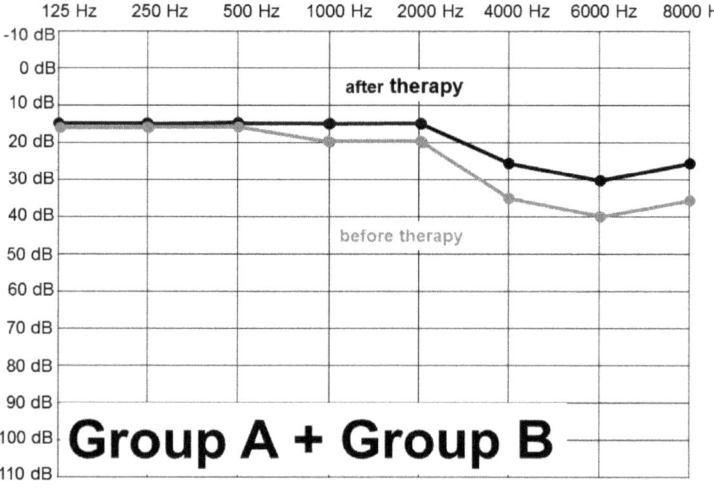

Figure 3. PTA before and after the treatment of AAT—in total in all AAT (n = 141 ears).

The success of particular medical protocols with/without HBO2 is referred to in Table 3. Improvement of hearing occurred in 82% of patients from Group I (without HBO2) and 77% of patients from Group II (with HBO2). A total of 134 damaged ears were evaluated. Group "Singled out group p. o. vasodilatants" treated with peroral vasodilatants, of which seven ears were not included—see Table 3.

Table 4 shows audiometric figures before and after the treatment in total in Group I (without HBO2) and Group II (with HBO2). The whole set in Group II shows a greater amelioration at most frequencies, a larger difference in the change of sound intensity was statistically significant at 500 Hz ($p < 0.01$) and 2000 Hz ($p < 0.05$). Cases of tinnitus occurred after AAT in 58% of 141 damaged ears. In Group A, tinnitus was experienced in 63% of ears and only 52% in Group B. After treatment, tinnitus disappeared in a total of 50% of damaged ears. In Group A, tinnitus disappeared in 54% of damaged ears compared to 43% in Group B.

Table 3. Success of particular medical protocols—Group I (corticosteroids therapy without HBO2), Group II (corticosteroids therapy with HBO2) and Singled out group (p. o. vasodilatants).

Group I (corticosteroids without HBO2)	n = 61 ears
Improved Improved to normacusis (threshold of losses above 20 dB)	50–82% 39–64%
Group II (corticosteroids with HBO2)	n = 73 ears
Improved Improved to normacusis (threshold of losses above 20 dB)	56–77% 27–37%
Singled out group p. o. vasodilatants	n = 7 ears
Improved Improved to standard (threshold of losses above 20 dB)	6–86% 2–29%

Table 4. Group I (corticosteroids therapy without HBO2) and Group II (corticosteroids therapy with HBO2) changes of threshold of hearing before and after the treatment.

Frequency	Group	N	Average (dB)	St. Deviation	*p*-Value *
125	I	61	1.48	4.117	0.984
	II	73	1.16	6.265	
250	I	61	1.39	3.180	0.121
	II	73	3.15	7.193	
500	I	61	1.23	3.248	0.007
	II	73	4.04	7.530	
1000	I	61	2.21	5.666	0.284
	II	73	3.63	8.261	
2000	I	61	3.11	5.490	0.043
	II	73	7.05	10.924	
4000	I	61	6.80	10.248	0.146
	II	73	10.41	12.984	
6000	I	61	9.18	16.985	0.767
	II	73	10.27	17.298	
8000	I	61	9.26	11.862	0.468
	II	73	8.56	16.083	

* Mann-Whitney test.

Vertigo occurred in 7% of patients from the total number of 108 patients (141 ears).

Nystagmus occurred in 3% of the total number of 108 patients. Vertigo and nystagmus disappeared after the treatment.

4. Discussion

Available sources state that, at present, AAT treatment consists of applying the combination of corticosteroid therapy and HBO2 therapy [1,3,4,11]. Studies proved the presence of steroid receptors in the inner ear. Steroids participate in forming ionic balance in the inner ear, stabilization of the cell membrane, and increased perfusion and inhibition of anti-inflammatory cytokines [4].

During HBO2 therapy, the patient inhales oxygen (100%) greater than at atmospheric pressure. This HBO2 therapy is generally administered at 2.0–2.8 atmospheres for a period of 60–90 min, usually once a day within 10–15 days [1–3]. HBO2 therapy increases the level of oxygen dissolved in the blood, which is subsequently transferred in larger amounts to tissues [1–3]. In our case, 2.5 atmospheres were utilized for 90 min.

The results acquired proved that improvement of hearing in patients with AAT is inversely proportional to the intensity of the damage. Thus, patients whose condition improved after the treatment had less severe hearing impairment before the start of the treatment in comparison with patients whose condition did not ameliorate after treatment—at 4000 Hz (at the significance level <0.05).

In terms of our results, it is possible to state, in addition, that hearing improvement depends on the age of the patient, thus patients who are statistically younger (average age of 34.4 years) see greater improvement in condition than older ones (average age of 44.4 years).

Lamm and Arnold demonstrated, on an animal model, the subsequent decrease of partial pressure in perilymph and the decrease of the amplitude of cochlear potentials during the first 24 min after AAT [4]. They, in turn, observed an immediate increase of these parameters, promptly after HBO2 therapy [4].

According to Reazee et al., the limit figure of the sound impulse causing AAT differs by various standards [19]. For instance, NATO set the safe sound threshold at 160 dB (for the army) [2]. The administration of safety and health protection at work considers 140 dB the safe limit of sound. In case of excessive exposure to noise exceeding this, the safe limit is further lowered [20,21]. Recently, an increased incidence of AAT has been a matter of discussion in connection with the use of Bren guns (a certain model of gun), which PTA displays damage prevalence between 3000–6000 Hz according to Mrena et al. [13] According to our data, the most severely damaged frequencies are 6000 Hz followed by 8000, 4000, and 2000 Hz respectively.

An interesting discovery in our dataset was found, as although we detected the success of the treatment lower in Group A to B (70 vs. 81%), there was a larger adjustment of hearing to normal in Group A over B (39 vs. 29%). Two possible influencing factors are considered—using new noisier Bren type guns and starting timely the treatment in Group A. For instance, Mardassi et al. [22] stated that improvement of hearing after treatment is 81% in a group of young soldiers (exposed to shots and explosions) (n = 64 ears). The hearing threshold improved, on average, by 14 dB. In our group, the amelioration after treatment was 8 dB. According to the Finnish study, it may be possible to prevent some AATs using careful planning of military exercises [12].

In addition, one of the most frequent symptoms in AAT is, excluding hypacusis and tinnitus. Its occurrence is caused by the disruption of functional compactness in hair cells and nerve fibers [2,4]. In the study by Jokitulppo et al., tinnitus after AAT occurred in more than 60%. Furthermore, the study states the exposure to noise correlates with subsequent tinnitus [12,23], which is evident in our group of patients, where the occurrence of tinnitus was similarly 58%.

Improvement of hearing depends on the age of the patient and several others additional comorbidities. The eventual presence of an endotympanic effusion must be evaluated and treated with a ventilation tube so that the auditory tube re-establishes its physiological function [24,25].

Van der Veen et al. in their study dealt with the clinical issue of how HBO2 therapy effects threshold figures of hearing in patients who suffered acute acoustic trauma [1]. They ascertained that the effect of HBO2 therapy on hearing thresholds in patients with hearing loss, caused by AAT, is not statistically significant. Thus, they recommend performing distinctly designed randomized controlled studies with a sufficient set of patients that could conclude HBO2's therapeutically effect in the treatment of AAT [1].

In our set, there were 52% ears in total, damaged by AAT, exposed to HBO2 therapy, and a positive effect was detected in 77%. On this basis, it may be possible to state that HBO2 therapy has a role in the treatment of AAT. A similar conclusion was given in a paper by Lafère et al. [26]. Also, Oya et al. published data on HBO2 therapy: 26 of the 37 ears (70%) displayed improved hearing [27].

In our study, we did not compare identical groups of patients (with HBO2 and without HBO2), because HBO2 therapy is indicated only in the case when pharmacological treatment is ineffective. Therefore, in accordance with Van der Veen et al., we recommend further studies so that it's conclusively possible to answer the question of the effectiveness of HBO therapy in AAT treatment [1].

5. Conclusions

In the Czech Republic, many patients with AAT are members of the armed forces and their trauma is often classified as an accident at work. Based on our study, we can confirm the positive effect of early initiated corticosteroid treatment and HBO2 therapy. The application of a treatment protocol, well-timed administration of corticosteroids and initiation of HBO2 (no later than 7 days after acoustic trauma) may help to improve hearing in a patient with AAT. Because many patients are military personnel who suffer from AAT after small arms fire, prevention and proper use of protective equipment are strongly recommended. Hyperbaric oxygen therapy shows to be an adjuvant option in AAT, unfortunately we did not get the same real evidence of effectiveness in recovery from concomitant tinnitus.

Author Contributions: Conceptualization, R.H. and J.A.; methodology, R.H., D.K. (Daniel Kovar) and P.D.; validation, R.H., K.P., P.D. and B.O.; investigation, S.Z., Z.F., K.J., and K.P.; data curation, S.Z., T.Y., and D.K. (David Kalfert); writing—original draft preparation, R.H., T.Y. and K.J.; writing—review and editing, D.K. (David Kalfert), Z.F., and J.A.; supervision, J.A.; project administration, R.H.; funding acquisition, J.A. All authors have read and agreed to the published version of the manuscript.

Funding: This work was supported by the Ministry of Defense of the Czech Republic—project MO 1012.

Institutional Review Board Statement: The study was conducted according to the guidelines of the Declaration of Helsinki.

Informed Consent Statement: Not applicable.

Data Availability Statement: Not applicable.

Acknowledgments: We highly appreciate the assistance of Alena Dohnalová, researcher of the Institute of Physiology of the 1st Faculty of Medicine of Charles University, who elaborated statistical data.

Conflicts of Interest: The authors declare no conflict of interest.

References

1. van der Veen, E.L.; van Hulst, R.A.; de Ru, J.A. Hyperbaric Oxygen Therapy in Acute Acoustic Trauma: A Rapid Systematic Review. *Otolaryngol. Head Neck Surg.* **2014**, *151*, 42–45. [CrossRef]
2. Baldwin, T.M. Tinnitus, a military epidemic: Is hyperbaric oxygen therapy the answer? *J. Spec. Oper. Med.* **2009**, *9*, 33–43. [PubMed]
3. Bayoumy, A.B.; van der Veen, E.L.; van Ooij, P.A.M.; Besseling-Hansen, F.S.; Koch, D.A.A.; Stegeman, I.; de Ru, J.A. Effect of hyperbaric oxygen therapy and corticosteroid therapy in military personnel with acute acoustic trauma. *J. R. Army Med Corps* **2019**. [CrossRef]
4. Lamm, K.; Arnold, W. The effect of prednisolone and non-steroidal anti-inflammatory agents on the normal and noise-damaged guinea pig inner ear. *Hear Res.* **1998**, *115*, 149–161. [CrossRef]
5. Quaranta, A.; Portalatini, P.; Henderson, D. Temporary and permanent threshold shift: An overview. *Scand. Audiol. Suppl.* **1998**, *48*, 75–86.
6. Cakir, B.O.; Ercan, I.; Civelek, S.; Körpinar, S.; Toklu, A.S.; Gedik, O.; Işik, G.; Sayin, I.; Turgut, S. Negative effect of immediate hyperbaric oxygen therapy in acute acoustic trauma. *Otol. Neurotol.* **2006**, *27*, 478–483. [CrossRef] [PubMed]
7. Colombari, G.C.; Rossato, M.; Feres, O.; Hyppolito, M.A. Effects of hyperbaric oxygen treatment on auditory hair cells after acute noise damage. *Eur. Arch. Otorhinolaryngol.* **2011**, *268*, 49–56. [CrossRef] [PubMed]
8. Kuznecov, M.S.; Morozova, M.V.; Dvorjanchikov, V.V.; Glaznikov, L.A.; Pastushenkov, V.L.; Gofman, V.R. Sovremennye podkhody i perspektivnye napravleniya v lechenii ostroi sensonevral'noi tugoukhosti akutravmaticheskogo geneza [Modern approaches and prospective directions in treatment of acute sensorineural hearing loss following acoustic trauma]. *Vestn. Otorinolaringol.* **2020**, *85*, 88–92. (In Russian) [CrossRef] [PubMed]
9. Kuokkanen, J.; Aarnisalo, A.A.; Ylikoski, J. Efficiency of hyperbaric oxygen therapy in experimental acute acoustic trauma from firearms. *Acta Otolaryngol. Suppl.* **2000**, *543*, 132–134. [CrossRef] [PubMed]
10. Holý, R.; Došel, P.; Synková, B.; Astl, J. Treatment of Idiopathic Sudden Sensorineural Hearing Loss—Hyperbaric Oxygen Therapy. *Otorinolaryngol. Foniatr.* **2017**, *66*, 135–140.
11. Ylikoski, J.; Mrena, R.; Makitie, A.; Kuokkanen, J.; Pirvola, U.; Savolainen, S. Hyperbaric oxygen therapy seems to enhance recovery from acute acoustic trauma. *Acta Otolaryngol.* **2008**, *128*, 1110–1115. [CrossRef]
12. Jokitulppo, J.; Toivonen, M.; Paakkonen, R.; Savolainen, S.; Bjork, E.; Lehtomaki, K. Military and leisure-time noise exposure and hearing thresholds of Finnish conscripts. *Mil. Med.* **2008**, *173*, 906–912. [CrossRef] [PubMed]

13. Mrena, R.; Savolainen, S.; Pirvola, U.; Ylikoski, J. Characteristics of acute acoustical trauma in the Finnish Defence Forces. *Int. J. Audiol.* **2004**, *43*, 177–181. [CrossRef] [PubMed]
14. Paakkonen, R.; Lehtomaki, K.; Savolainen, S. Noise attenuation of communication hearing protectors against impulses from assault rifle. *Mil. Med.* **1998**, *163*, 40–43. [CrossRef] [PubMed]
15. Ylikoski, M.E. Prolonged exposure to gunfire noise among professional soldiers. *Scand. J. Work Environ. Health* **1994**, *20*, 87–92. [CrossRef]
16. Kratochvilova, B.; Profant, O.; Astl, J.; Holy, R. Our experience in the treatment of idiopathic sensorineural hearing loss (ISNHL): Effect of combination therapy with HBO(2) and vasodilator infusion therapy. *Undersea Hyperb. Med.* **2016**, *43*, 771–780.
17. Ciodaro, F.; Gazia, F.; Galletti, B.; Galletti, F. Hyperbaric oxygen therapy in a case of cervical abscess extending to anterior, mediastinum, with isolation of *Prevotella corporis*. *BMJ Case Rep.* **2019**, *12*, e229873. [CrossRef]
18. Ferlito, S.; Maniaci, A.; Di Luca, M.; Grillo, C.; Mannelli, L.; Salvatore, M.; La Mantia, I.; Spinato, G.; Cocuzza, S. From Uncommon Infection to Multi-Cranial Palsy: Malignant External Otitis Insights. *Dose Response* **2020**, *18*, 1559325820963910. [CrossRef]
19. Rezaee, M.; Mojtahed, M.; Ghasemi, M.; Saedi, B. Assessment of impulse noise level and acoustic trauma in military personnel. *Trauma Mon.* **2012**, *16*, 182–187. [CrossRef]
20. Olszewski, J.; Milonski, J.; Olszewski, S.; Majak, J. Hearing threshold shift measured by otoacoustic emissions after shooting noise exposure in soldiers using hearing protectors. *Otolaryngol. Head Neck Surg.* **2007**, *136*, 78–81. [CrossRef]
21. Tambs, K.; Hoffman, H.J.; Borchgrevink, H.M.; Holmen, J.; Engdahl, B. Hearing loss induced by occupational and impulse noise: Results on threshold shifts by frequencies, age and gender from the Nord-Trondelag Hearing Loss Study. *Int. J. Audiol.* **2006**, *45*, 309–317. [CrossRef]
22. Mardassi, A.; Turki, S.; Mbarek, H.; Hachicha, A.; Benzarti, S.; Abouda, M. Acute acoustic trauma: How to manage and how to prevent? *Tunis. Med.* **2016**, *94*, 664. [PubMed]
23. Tlapak, J.; Chmatal, P.; Oniscenko, B.; Pavlik, V.; Dosel, P.; Paral, J.; Lochman, P. The effect of hyperbaric oxygen therapy on gene expression: Microarray analysis on wound healing. *Undersea Hyperb. Med.* **2020**, *47*, 31–37. [CrossRef] [PubMed]
24. Ferlito, S.; Cocuzza, S.; Grillo, C.; La Mantia, I.; Gulino, A.; Galletti, B.; Coco, S.; Renna, C.; Cipolla, F.; Di Luca, M.; et al. Complications and sequelae following tympanostomy tube placement in children with effusion otitis media: Single center experience and review of literature. *Acta Med. Mediterr.* **2020**, *36*, 1905–1912.
25. Galletti, F.; Freni, F.; Gazia, F.; Galletti, B. Endomeatal approach in cochlear implant surgery in a patient with small mastoid cavity and procident lateral sinus. *BMJ Case Rep.* **2019**, *12*, e229518. [CrossRef]
26. Lafère, P.; Vanhoutte, D.; Germonprè, P. Hyperbaric oxygen therapy for acute noise-induced hearing loss: Evaluation of different treatment regimens. *Diving Hyperb. Med.* **2010**, *40*, 63–67. [PubMed]
27. Oya, M.; Tadano, Y.; Takihata, Y.; Ikomi, F.; Tokunaga, T. Utility of Hyperbaric Oxygen Therapy for Acute Acoustic Trauma: 20 years' Experience at the Japan Maritime Self-Defense Force Undersea Medical Center. *Int. Arch. Otorhinolaryngol.* **2019**, *23*, e408–e414. [CrossRef]

International Journal of
Environmental Research and Public Health

MDPI

Article

Influence of Recurrent Laryngeal Nerve Transient Unilateral Palsy on Objective Voice Parameters and on Voice Handicap Index after Total Thyroidectomy (Including Thyroid Carcinoma)

Zuzana Veldova [1,2], Richard Holy [1,2,*], Jan Rotnagl [1,2], Temoore Younus [1,2], Jiri Hlozek [1,2] and Jaromir Astl [1,2]

[1] Department of Otorhinolaryngology and Maxillofacial Surgery, Military University Hospital, 16902 Prague, Czech Republic; zuzana.veldova@uvn.cz (Z.V.); jan.rotnagl@uvn.cz (J.R.); temoore.younus@lf3.cuni.cz (T.Y.); jiri.hlozek@uvn.cz (J.H.); jaromir.astl@seznam.cz (J.A.)

[2] Third Faculty of Medicine, Charles University, 10000 Prague, Czech Republic

* Correspondence: richard.holy@uvn.cz

Citation: Veldova, Z.; Holy, R.; Rotnagl, J.; Younus, T.; Hlozek, J.; Astl, J. Influence of Recurrent Laryngeal Nerve Transient Unilateral Palsy on Objective Voice Parameters and on Voice Handicap Index after Total Thyroidectomy (Including Thyroid Carcinoma). *Int. J. Environ. Res. Public Health* **2021**, *18*, 4300. https://doi.org/10.3390/ijerph18084300

Academic Editors: Francesco Galletti, Francesco Gazia, Francesco Freni, Cosimo Galletti and Anna M. Giudetti

Received: 23 March 2021
Accepted: 16 April 2021
Published: 18 April 2021

Publisher's Note: MDPI stays neutral with regard to jurisdictional claims in published maps and institutional affiliations.

Abstract: *Introduction:* Total thyroidectomy (TT) is one of the most common surgical endocrine surgeries. Voice impairment after TT can occur not only in patients with recurrent laryngeal nerve (RLN) transient paralysis, but also in cases of normal vocal cord mobility. *Aim:* To compare voice limits using a speech range profile (SRP) in patients before and 14 days after TT and to investigate the influence of the early results of voice quality after TT on the personal lives of patients. We focused on the perception of voice change before and shortly after TT. *Materials and methods:* A retrospective study, in the period 2018–2020, included 65 patients aged 22–75 years. We compared two groups of patients: group I (n = 45) (without RLN paresis) and group II (n = 20) (with early transient postoperative RLN paresis). Patients underwent video flexible laryngocopy, SRP, and Voice Handicap Index-30 (VHI-30). *Results:* In group I, the mean values of F_{max} (maximum frequency) and I_{max} (maximum intensity) decreased in women (both $p = 0.001$), and VHI-30 increased ($p = 0.001$). In group II after TT in women, the mean F_{max} and I_{max} values decreased ($p = 0.005$ and $p = 0.034$), and the frequency range of the voice was reduced from 5 to 2 semitones. The dynamic range of the voice was reduced by 3.4 dB in women and 5.1 dB in men.VHI-30 increased ($p = 0.001$). *Conclusion:* The study documented a worsening of the mean values of SRP, VHI-30, and voice parameters of patients in group II. Voice disorders also occurred in group I without RLN paresis. Non-paretic causes can also contribute to voice damage after TT. SRP and VHI-30 are suitable tools for comparing voice status in two groups of patients, including those with dysphonia. Our data support the claim that the diagnosis of a thyroid cancer does not necessarily imply a higher postoperative risk of impaired voice quality for the patient.

Keywords: total thyroidectomy; recurrent laryngeal nerve paresis; Voice Handicap Index; speech range profile

1. Introduction

Voice changes after total thyroidectomy (TT) often have a neurogenic cause and are one of the most common complications after thyroid surgery. TT is indicated for the treatment of autoimmune inflammatory disorders (thyrotoxicosis), endemic goiter, and malignant and benign tumors. The change in voice in patients after TT can be caused by

- nerve injury during TT;
- paretic (or plegic) causes of voice damage: recurrent laryngeal nerve (RLN) paresis or superior laryngeal nerve (SLN) paresis;
- a non-paretic cause (e.g., laryngopharyngeal reflux disease (LPRD)).

Only a thorough videolaryngostroboscopy examination performed in the preoperative period can reveal the above causes of dysphonia. A patient's vocal cords must be examined before each TT. These must be differentiated from potential laryngeal nerve injury-RLN paresis after TT [1]. Hoarseness caused by RLN paresis is one of the most common postoperative complications of TT after hypocalcaemia [2]. Consequences resulting from nerve injury voice deterioration after TT (postoperative RLN paralysis) differs depending on whether the injury of RLN corresponds to neurapraxia (nerve compression, which primarily damages the myelin sheath and secondarily the axons), axonotmesis (traction injury by excessive nerve strain followed by axon fracture) or neurotmesis (complete or partial nerve break) according to Seddon's Classification [3–5] and whether the damage or injury is unilateral or bilateral.

The incidence of postoperative RLN paresis ranges from 0–8% in primary thyroid surgery. It is used as a criterion to measure the success of surgical treatment [6].

The identification of RLN injuries in terms of transient paresis in 0.7% of patients, and persistent paresis in 0.2% of patients, is reported by some authors [7].

RLN injury during thyroidectomy/parathyroidectomy occurs intraoperatively significantly more frequently in visually intact RLN than in the transected nerves [8]. The forward motor branch of RLN bifurcation near the Berry ligament is particularly at risk of traction injury. This is made evident by a deterioration in voice quality after TT (higher average jitter) in an effort to maintain the radicality of surgery (Berry ligament) in the cases of malignant thyroid tumors [6]. TT is a standard treatment for thyroid tumors including carcinomas [7,9].

Voice change after TT is reported in the literature in 16–89% of cases [10,11], but is often not considered to be associated with neurogenic impairment (RLN paresis) [10,12].

Suitable questionnaires for the subjective evaluation of VHI voice quality are reported in the literature [10,13,14]. In practice, the VHI-10 and VHI-30 questionnaires are used as an evaluation of voice quality as a parameter of postoperative quality of life.

There is an Italian study of voice professionals (teachers) that shows the benefits of the VoiSS VTDS questionnaires for a preventive voice program for teachers [15].

The non-paretic causes of voice change after TT include

- aero-digestive disorders: proximal acid reflux–loss of coordination of upper esophageal sphincter (UES) [10];
- perioperative trauma of extralaryngeal muscles [11,16];
- modified vascular supply and venous drainage-postoperative soft tissue edema, and mucosal edema-postintubation (orotracheal) [9];
- local neck pain (psychological responses after surgery) [12];
- changes in laryngeal mucosa;
- neurogenic (psychogenic) causes.

Patients without RLN paresis may have problems with voice quality, early voice fatigue, and limited voice range. Regardless of the etiology of voice disorders after TT, voice change is a feared complication for patients, especially vocal professionals (opera singers, pop music singers, actors, teachers, speakers, managers, translators, members of choirs, actresses, lawyers, doctors). That is why follow-up phoniatric care is so important for all patients who have undergone TT. Logopedic therapy in patients with post-thyroidectomy dysphonia improves the vocal and postural outcomes [14].

2. Materials and Methods

Our monocentric study included 65 patients indicated for TT. All patients were examined between January 2018 and September 2020. The group consisted of 57 women and 8 men, and the mean age in the group was 45 years (range 22–75 years).

All operations were performed by senior surgeons at one institution.

During all TT operations intraoperative neuro-monitoring (IONM) was used.

In the case of IONM, short- or medium-acting muscle relaxants were used on administration of anesthesia. Invasive needle sensing electrodes were applied to the m. vocalis.

Non-invasive electrodes present on the oro-tracheal tube were placed between the voice ligaments under the control of direct laryngoscopy.

A monopolar stimulation electrode was used (stimulus intensity 0.5–1 mA). We evaluated action potential (AP) with an amplitude of more than 100 µV with adequate latency from the stimulation site as an adequate functional nerve response.

Electro-myographic recording was performed on a MEDTRONIC instrument NIM-NEURO 3.0. A true-positive result (TP) occurred when the IONM detected RLN palsy, which was confirmed postoperatively.

A false positive was considered to be a condition where, despite the IONM anticipated RLN paresis, the vocal cords were postoperatively moving. The true negative result was a condition where the IONM assumed that the RLN functionality was maintained, which was confirmed postoperatively. False negative results were evaluated as a condition where, despite the IONM detecting preserved function of RLN, postoperative paresis occurred.

To reduce the incidence of false-negative and false-positive responses, we monitored neural structures according to the scheme: vagus nerve before dissection, recurrent laryngeal nerve before dissection, RLN after dissection, vagus nerve after dissection.

All patients included in the study (with intraoperative IONM obtained in a non-physiological/pathological response or postoperatively caused by changing voice) were promptly investigated (objective voice parameters and VHI-30 questionnaire) within 14 days after TT.

The inclusion criteria were as follows:

- age over 18 years;
- preoperative normal laryngeal finding;
- preoperatively normal voice;
- non-physiological/pathological response obtained perioperatively at IONM;
- postoperative change of voice.

The exclusion criteria were as follows:

- age below 18 years;
- history of any neurological disorders or diseases (e.g., multiple sclerosis);
- any preoperative benign vocal cord lesions or other voice disorders;
- any preoperative pathological videolaryngoscopic findings;
- hearing loss requiring hearing aids.

We divided the patients after TT into two groups:

- GROUP I with movable vocal cords (verified by laryngostroboscopic examination) (without RLN paresis) $n = 45$;
- GROUP II patients with postoperative unilateral transient RLN paresis $n = 20$.

The indications for TT surgery were as follows: nodular goiter $n = 25$ (38%), thyroiditis with thyreotoxicosis $n = 10$ (15%), diffuse goiter $n = 7$ (11%), Graves–Basedow goiter n = 6 (9%), multinodular goiter $n = 6$ (9%), papillary thyroid carcinoma (papillary carcinoma, papillary variant of papillary carcinoma, foliculary variant of papillary carcinoma) $n = 6$ (9%), Hashimoto´s autoimmune thyroiditis $n = 3$ (5%), medullary thyroid carcinoma $n = 1$ (2%), and retrosternal nodular goiter $n = 1$ (2%).

All patients underwent the following examinations before TT and check-up within 14 days after TT.

Performed examinations:

- Optical methods: video flexible laryngoscopy;
- Acoustic methods: jitter (%);
- Examination of voice field of conversational voice (determination of basic average voice position Fo (Hz) and basic sound pressure level SPL (dB (A));
- Aerodynamic examination: maximum phonation time (MPT) (s);
- Psychometric examination: VHI-30 (Voice Handicap Index);
- Speech range profile (SRP) (dynamic and frequency range of speaking voice) loud reading and reading with low voice intensity not whisper.

We used the LingWAVES software system for voice analysis.

We analyzed the periodicity of the voice using the perturbation parameter (jitter) and examined the voice field of the speech voice (determination of the average mean voice position Fo (Hz) and the basic sound pressure level SPL (dB (A))), SRP (speech range profile) using silent and loud reading methods to determine the lowest frequency (F_{min} Hz), highest frequency (F_{max} Hz), minimum intensity (I_{min} dB SPL), and maximum intensity (I_{max} dB SPL). We evaluated the changes in the glottal gap using an aerodynamic test using the prolonged phonation of the isolated vowel (A), thus measuring the maximum phonation time. All sound recordings were made at a distance of 30 cm from the microphone in a quiet room with an ambient noise level of less than 40 dB (A)). When recording standard text readings, the normal volumes of the conversational voice (habitual reading), silent voice (semi-voice), and loud reading were used.

Psychometric examinations were evaluated by completing the VHI-30 (Voice Handicap Index) questionnaires.

Methodology: All patients signed an informed consent for total thyroidectomy before the surgeries. Our study was followed according to the Declaration of Helsinki. Due to the retrospective type and character of the study, approval by the local ethics committee was not required.

3. Results

The total percentage of transient RLN paresis was 31%. It was always a transient RLN paresis.

From the point of view of thyroid histology, transient unilateral paresis was most often found in group II in patients with nodular goiter, $n = 8$ (40%), and thyreotoxicosis, $n = 5$ (25%).

Laterality of transient RLN Paresis in Group II:

In terms of laterality, transient RLN paresis prevailed on the left side: 60% ($N = 12$). Right-sided transient RLN paresis was 40% ($N = 8$).

Results of objective voice parameters:

In Group I, only 2 parameters changed significantly after the TT: jitter increase ($p = 0.001$), and VHI-30 point increase ($p = 0.001$).

Fo values in women (habitual voice-reading standard text with medium voice intensity) were statistically significantly lower than before the operation ($p = 0.016$).

The values of Fo in men did not differ statistically significantly after TT; there was only a slight deepening of the voice (decrease in the pitch).

In Group II after the TT, all observed parameters changed significantly (worsened): MPT shortening ($p = 0.001$), SPL decrease ($p = 0.030$), jitter increase ($p = 0.001$), and VHI-30 point increase ($p = 0.001$).

The values of Fo (habitual voice-reading standard text with medium voice intensity) in women did not change significantly after TT ($p = 0.836$).

The values of Fo in men differed only on the significance level $p = 0.1$ (not statistically significant); we noticed a slight increase in voice position and an increase in voice pitch.

Test speech range profile (SRP) (the most important parameter) results are shown in see Table 1.

SRP Group I:

Basic statistical characteristics of $F_{min}/_{max}$ and $I_{min}/_{max}$ and the results of the non-parametric Wilcoxon test between before and after TT (before and after):

Women: the values for loud reading differed statistically significant: for F_{max}, voice height decreased after surgery ($p = 0.001$). But I_{max} also decreased after surgery, there was a limitatiton of load reading ability ($p = 0.001$).

Men: none of the monitored parameters proved to be statistically significantly different between F_{max} and I_{max} after TTE. The pitch of the voice or the intensity of the voice in the upper part of the voice field (VF) were almost unchanged.

Table 1. SRP: mean values, standard deviations, statistical significance.

Parameter	*n*	Before	After	*p*
Group I women				
F_{min} (Hz)	41	195.45 ± 70.22	194.59 ± 74.22	0.588
F_{max} (Hz)	41	256.09 ± 84.29	242.96 ± 87.62	0.001
I_{min} (dB (A))	41	59.64 ± 19.66	59.64 ± 19.30	0.423
I_{max} (dB (A))	41	74.74 ± 22.59	72.60 ± 21.59	0.001
Group I men				
F_{min} (Hz)	4	127.35 ± 42.60	111.13 ± 18.03	0.655
F_{max} (Hz)	4	177.75 ± 60.16	172.25 ± 64.69	0.655
I_{min} (dB (A))	4	56.23 ± 3.93	57.13 ± 2.54	0.285
I_{max} (dB (A))	4	68.55 ± 4.51	70.73 ± 5.55	0.285
Group II women				
F_{min} (Hz)	16	190.27 ± 24.21	193.43 ± 22.08	0.328
F_{max} (Hz)	16	249.57 ± 31.49	228.26 ± 18.73	0.005
I_{min} (dB (A))	16	54.54 ± 2.74	54.41 ± 3.71	0.477
I_{max} (dB (A))	16	68.74 ± 3.99	65.17 ± 6.59	0.034
Group II men				
F_{min} (Hz)	4	118.93 ± 35.25	118.48 ± 31.33	1.000
F_{max} (Hz)	4	153.45 ± 36.70	140.85 ± 38.95	0.109
I_{min} (dB (A))	4	51.58 ± 4.45	51.08 ± 1.65	1.000
I_{max} (dB (A))	4	70.60 ± 2.37	64.95 ± 3.31	0.068

SRP = speech range profile; F_{min} = minimum frequency; F_{max} = maximum fraquency; I_{min} = minimum intensity; I_{max} = maximum intensity.

SRP Group II:

Basic statistical characteristics of $F_{min}/_{max}$ and $I_{min}/_{max}$ and the results of the non-parametric Wilcoxon test before and after TTE:

Women: after TT, the VF (voice field) parameters of the loud reading F_{max} and I_{max}, were significantly different ($p = 0.005$) and ($p = 0.034$). Here, too, there was a reduction in voice height while trying to read loudly, thus limiting the ability to amplify the voice.

Men: F_{max} values did not change after TT; only I_{max} differed, but only at a significance level of 0.1 ($p = 0.068$). Thus, the ability to read aloud in men with RLNp after TT was not significantly reduced.

Results—voice field speaking voice SRP (see Table 2):

Table 2. Voice field of the speaking voice.

Female	Group I		Group II	
FR (Hz) before TT (Hz) $_{(max-min)}$	60.7 (G3-C4)	5 st	59.3 (F3#-H3)	5 st
FR (Hz) after TT (Hz) $_{(max-min)}$	48.2 (G3-H3)	4 st	34.8 (G3-A3)	2 st
DR dB SPL (dB (A)) before TT $_{(max-min)}$	15.1		14.2	
DR dB SPL (dB (A)) after TT $_{(max-min)}$	13.0		10.8	
Difference FR (before/after TT) Difference DR (before/after TT)	−1 st −2.1 dB (SPL(A))		−3 st −3.4 dB (SPL(A))	
Male	Group I		Group II	
FR (Hz) before TT $_{(max-min)}$	50.4 (C3-F3)	5 st	35.5 (A2#-D3#)	5 st
FR (Hz) after TT $_{(max-min)}$	61.1 (A2-F3)	8 st	22.4 (A2-C3#)	4 st
DR dB SPL (dB (A)) before TT $_{(max-min)}$	12.3		19.0	
DR dB SPL (dB (A)) after TT $_{(max-min)}$	13.6		13.9	
Difference FR (before/after TT) Difference DR (before/after TT)	+3 st +1.3 dB (SPL(A))		−1 st −5.1 dB (SPL(A))	

FR = frequency range; DR = dynamic range; F_{min} = minimum frequency; F_{max} = maximum frequency; I_{min} = minimum intensity; I_{max} = maximum intensity; st = number of semitones; # = semitone higher than previous full tone.

Group I Women: The voice frequency range from the original 5 semitones (G3-C4) was limited to 4 semitones (G3-H3) in the upper part of the voice field. Dynamic voice range reduced from 15.1 dB SPL (dB (A)) to 13.0 dB SPL (dB (A)) mainly at higher voice field frequencies (limiting voice intensity during loud reading).

Men: After the TT, the lower part of the voice field (VF) was freely expanded and the upper part of the VF in loud reading remained unchanged. There was a change in the extent of VF before surgery 5 semitones (C3-F3) versus after surgery 8 semitones (A2-F3).

The dynamic range of the voice increased at loud reading, at the same frequency level as before TT, i.e., 12.3 dB (SPL (A)) before and 13.6 dB (SPL (A)) after TT.

Group II Women: We noticed the restriction in the voice field in the frequency and dynamic range, in both parts of the voice field and in the lower and upper parts.

The voice pitch was limited in both the low and high frequencies. Frequency range decreased from 5 semitones (F3#-H3) to 2 semitones (G3-A3#); voice dynamics decreased from 14.2 dB SPL (dB (A)) to 10.7 dB SPL (dB (A)).

Men: Shifting voice pitch to deeper parts of the frequency spectrum, limited ability to increase voice pitch and intensity from loud reading, from original 4 semitones (A2#-D3#) to 3 semitones (A2#-C3#), dynamic range reduced from 19.0 dB SPL (dB (A)) to 13.9 dB SPL (dB (A)).

Results of VHI-30 questionnaires (see Table 3):

Table 3. Voice Handicap Index (VHI) VHI–30.

	Group I $n = 45$		**Group II** $n = 20$	
Part of the VHI-30	**Before TT Average Value**	**After TT Average Value**	**Before TT Average Value**	**After TT Average Value**
Physical	0.91	3.7	0.85	7.9
Functional	0.37	2.3	1.0	6.3
Emotional	0.24	1.7	0.4	3.7
VHI-Total (SD)	1.6 (2.99)	6.93 (5.95)	2.2 (2.42)	18.4 (19.22)

SD = Standard Deviation.

The total score significantly changed (worsened) in Group II ($p = 0.001$), but also in Group I ($p = 0.001$).

Our results: We noticed mild voice problems when evaluating the each parts of the VHI-30 (Physical, Functional, Emotional parts), VHI-30 Total—we noticed no patient deteriorated by more than 18 points. Overview in Table 3.

Patients with thyroid Carcinoma:

5 patients with papillary carcinoma were without paresis postoperatively.

1 patient with papillary carcinoma and 1 patient with medullary carcinoma had transient unilateral RLN paresis.

In the group of 5 patients with thyroid Carcinoma without postoperative RLN paresis, there was a reduction in FR by an average of 1 semitone. VHI-30 total points increased by only 5.5 points. The dynamic range of the voice decreased after TTE by 4 dB SPL (dB (A)), by 0.64 dB SPL (dB (A)) more than the average in the group of all patients without postoperative RLN paresis.

1 patient with papillary carcinoma had transient unilateral RLN paresis. After TTE, the VHI-30 increased from 6 to 70 points. The dynamic range of the voice after TTE decreased by 13 dB SPL (dB (A)).

1 patient with medullary carcinoma had transient unilateral RLN paresis. After TTE, the VHI-30 had not changed before 4/after 4 points. The dynamic range of the voice after TTE decreased by only 1.3 dB SPL (dB (A)). This patient had not been shown to have NEM2A or B, Hirschsprung's disease.

4. Discussion

The results of our study indicate, in agreement with the world literature, the importance of two parameters in monitoring voice analysis in patients after TT [17]: SRP (speech range profile) and VHI-30. Previously monitored parameters such as jitter and shimmer are losing importance in voice quality assessment today.

Our results support the idea that the SRP is an important indicator of the change in voice after TT. Our data show a difference of Fo max–Fo min in women in group II (with RLN paresis) during loud reading and semi-voice reading, limiting the frequency range of the voice from the original 5 semitones (F3#-H3) to 2 semitones (G3-A3), especially in the higher frequency range. The finding correlates, among other things, with the finding of glottis insufficiency due to unilateral RLN paresis after TT. In contrast, Bihari et al. found in their study that there was no change in tone range in patients with unilateral RLN paresis [18].

Reduction of the frequency range of the voice in women of Group II in our study, of 3 semitones in both the lower and upper parts of the frequency spectrum, may be due to both paretic (RLNparesis, EBSLNp) and non-paretic causes of voice disorders (most often proximal acid reflux or mucosal-orotracheal postintubation edema).

Another part of the SRP rating is Dynamic Voice Range and it has a similar predicative value. Our results show that Group II women with RLN paresis were able to use the same low voice intensity (I_{min}) after TT as before surgery, which was measured during reading by semi-voice.

According to some authors such as Ma et al. and Siupsinskiene et al. [19,20], the lower the I_{min}, the lower the subglottic pressure. The I_{min} parameter is a basic predictor of voice disorder, which is one of the basic indicators of improving the dynamic range of the voice during rehabilitation.

According to Leino et al., in men, an enormous increase in I_{max} is a sign of voice fatigue, and at the same time a decrease in the dynamic range of the voice signals a voice disorder [21].

In our study in Group I (without RLN), paradoxically, after the TT, the dynamic range of the speaking voice was expanded.

In Group II (with RLN paresis), however, as expected, the Dynamic Voice Range was reduced by 5.1 dB SPL (dB (A)).

Our study shows that I_{max} is not, in contrast to Leino et al. [21], a significant indicator of voice disorders in men (after TT). With regard to changes in the dynamic range of voice (decrease) our results are similar to Leino et al. [21].

D´Alatri with Marchese showed in their research the importance of the SRP.

They compared SRP and VRP (Voice Range Profile) in healthy subjects and dysphonic patients [22].

In common phoniatric practice, VRP (singing examination) is performed only on vocal professionals. VRP has no application in patients with dysphonia that has occurred after TT. The SRP is a useful alternative tool for assessing voice limits [22].

The importance of VHI-30-evaluation of individual parts of VHI-30 questionnaires after TT shows an increase in the number of points (deterioration) in the physical, functional, and emotional parts. In both of our evaluated groups, the most dissatisfaction with voice was manifested in the physical part, and the least in the emotional part.

Our results correlate with Dehqan et al. [23], where the authors reported in their group of patients with unilateral RLN paresis a direct correlation between VHI-30 total and the physical part of VHI-30 [23]. In Group I (without RLN transient paresis), a worsening of the total VHI-30 score was found in 31% of patients, whereas in Group II (with transient RLN paresis), the worsening of the total number was 75%.

In our study, in Groups I and II we noticed a direct correlation between the subjective assessment of patient voice quality and with the results of their voice analysis parameters and SRP (frequency and dynamic range of voice). Patients with the greatest SRP restriction and impaired perturbation parameters were more dissatisfied with their voice and vice versa.

Although these included patients without RLN paresis, they had worse VHI-30 scores. This fulfills our assumptions that non-paretic causes of voice changes (alongside paretic causes) and predictive factors contribute to the resulting patient satisfaction with their voice and to their application in personal and professional life.

The results of the SRP and VHI-30 examinations are relevant for the future, as they contribute to the development of new rehabilitation procedures, which will enable patients to use their voice without restrictions in communication processes in everyday life.

5. Conclusions

The study shows that in the case of monitoring changes in voice after TT, the most important objective voice parameters are SRP and VHI-30. These are very sensitive indicators of voice changes that can be used both in patients with dysphonia (after TT) caused by transient RLN paresis and in patients with non-paretic dysphonia.

Our data support the claim that the diagnosis of thyroid cancer does not necessarily imply a higher postoperative risk of impaired voice quality for the patient.

Author Contributions: Conceptualization, Z.V. and R.H.; methodology, Z.V. and R.H.; software, J.H. and T.Y.; validation, J.R., Z.V. and T.Y.; formal analysis, J.R.; investigation, Z.V. and R.H.; resources, J.H. and J.A.; data curation, J.H. and J.R.; writing—original draft preparation, Z.V. and R.H.; writing—review and editing, J.A. and R.H.; visualization, J.R. and J.H.; supervision, J.A.; project administration, R.H.; funding acquisition, J.A. All authors have read and agreed to the published version of the manuscript.

Funding: This research was funded by Ministry of the Ministry of Defense of the Czech Republic, Project MO 1012.

Institutional Review Board Statement: The retrospective study was conducted according to the guidelines of the Declaration of Helsinki. Due to the retrospective type and character of the study, approval by the local ethics committee was not required.

Informed Consent Statement: Informed consent was obtained from all subjects involved in the study.

Data Availability Statement: Not applicable.

Conflicts of Interest: The authors declare no conflict of interest.

References

1. Nam, I.C.; Bae, J.S.; Shim, M.R.; Hwang, Y.S.; Kim, M.S.; Sun, D.I. The Importance of Preoperative Laryngeal Examination Before Thyroidectomy and the Usefulness of a Voice Questionnaire in Screening. *World J. Surg.* **2012**, *36*, 303–309. [CrossRef] [PubMed]
2. Bhattacharyya, N.; Fried, M.P. Assessment of the Morbidity and Complications of Total Thyroidectomy. *Arch. Otolaryngol. Head Neck Surg.* **2002**, *128*, 389–392. [CrossRef] [PubMed]
3. Ambler, Z. Neuropatie a myopatie. In *Neuropathy and Myopathy*, 1st ed.; Triton, Ltd.: Prague, Czech Republic, 1999.
4. Chhabra, A.; Ahlawat, S.; Belzberg, A.; Andreseik, G. Peripheral nerve injury grading simplified on MR neurography: As referenced to Seddon and Sunderland classifications. *Indian J. Radiol. Imaging* **2014**, *24*, 217–224. [CrossRef] [PubMed]
5. Sunderland, S. The anatomy and physiology of nerve injury. *Muscle Nerve* **1990**, *13*, 771–784. [CrossRef] [PubMed]
6. Astl, J. Chirurgická léčba nemocí štítné žlázy 2. rozšířené vydání. In *Surgical Treatment of Thyroid Diseases*, 2nd ed.; Maxdorf, Ltd.: Prague, Czech Republic, 2014.
7. Lee, Y.S.; Nam, K.H.; Chung, W.Y.; Chang, H.S.; Park, C.S. Postoperative Complications of Thyroid Cancer in a Single Center Experience. *J. Korean Med. Sci.* **2010**, *25*, 541–545. [CrossRef]
8. Snyder, S.K.; Lairmore, T.C.; Hendricks, J.C.; Roberts, J.W. Elucidating Mechanisms of Recurrent Laryngeal Nerve Injury During Thyroidectomy and Parathyroidectomy. *J. Am. Coll. Surg.* **2008**, *206*, 123–130. [CrossRef]
9. Maeda, T.; Saito, M.; Otsuki, N.; Morimoto, K.; Takahashi, M.; Iwaki, S.; Inoue, H.; Tomoda, C.; Miyauchi, A.; Nibu, K.-I. Voice Quality after surgical treatment for thyroid cancer. *Thyroid* **2013**, *23*, 847–853. [CrossRef]
10. Svec, J.; Lejska, M.; Frostova, J.; Zabrodsky, M.; Drsata, J.; Kral, P. Česká verze dotazníku Voice Handicap Index pro kvantitativní hodnocení hlasových potíží vnímaných pacientem. (Version of the Voice Handicap Index Questionnaire for Quantitative Evaluation of Voice Problems Percieved by Patiens.) Otorinolaringologie a foniatrie. *Praha* **2009**, *58*, 132–139.

11. Henry, L.R.; Solomon, N.P.; Howard, R.; Gurevich-Uvena, J.; Horst, L.B.; Coppit, G.; Orlikoff, R.; Libutti, S.K.; Shaha, A.R.; Stojadinovic, A. The functional impact on voice of sternothyroid muscle division during thyroidectomy. *Ann. Surg. Oncol.* **2008**, *15*, 2027. [CrossRef]
12. Sinagra, D.L.; Montesinos, M.R.; Tacchi, V.A.; Moreno, J.E.; Falco, J.E.; Mezzadri, N.A.; DeBonis, D.L.; Curutchet, H. Voice changes after thyroidectomy without recurrent laryngeal nerve injury. *J. Am. Coll. Surg.* **2004**, *199*, 556–560. [CrossRef]
13. Rossi, V.C.; Fernandes, F.L.; Ferreira, M.A.; Bento, L.R.; Pereira, P.S.; Chone, C.T. Larynx cancer: Quality of life and voiceafter treatment. *Braz. J. Otorhinolaryngol.* **2014**, *80*, 403–408. [CrossRef]
14. Bruno, G.; Melissa, S.; Natalia, C.; Francesco, G.; Francesco, F.; Rocco, B.; Patrizia, L.; Antonella, P.; Ettore, C.; Zhang, D.; et al. Posture and dysphonia associations in patients undergoing total thyroidectomy: Stabilometric analysis. *Updates Surg.* **2020**, *72*, 1143–1149. [CrossRef]
15. Galletti, B.; Sireci, F.; Mollica, R.; Iacona, E.; Freni, F.; Martines, F.; Scherdel, E.P.; Bruno, R.; Longo, P.; Galletti, F. Vocal Tract Discomfort Scale (VTDS) and Voice Symptom Scale (VoiSS) in the Early Identification of Italian Teachers with Voice Disorders. *Int. Arch. Otorhinolaryngol.* **2020**, *24*, e323–e329. [CrossRef]
16. Cohen, S.M.; Jacobson, B.H.; Garett, C.G.; Noordzij, J.P.; Stewart, M.G.; Attia, A.; Ossoff, R.H.; Cleveland, T.F. Creation and validation of the Singing Voice Handicap Index. *Ann. Otol. Rhinol. Laryngol.* **2007**, *116*, 402–406. [CrossRef]
17. Lee, D.Y.; Lee, K.J.; Hwang, S.M.; Oh, K.H.; Cho, J.G.; Baek, S.K.; Kwon, S.-Y.; Woo, J.-S.; Jung, K.-Y. Analysis of Temporal Change in Voice Quality After Thyroidectomy: Single-institution Prospective Study. *J. Voice* **2017**, *31*, 195–201. [CrossRef]
18. Bihari, A.; Meszaros, K.; Remenyi, A.; Lichtenberger, G. Voice quality improvement after management of unilateral vocal cord paralysis with different techniques. *Eur. Arch. Otorhinolaryngol.* **2006**, *263*, 1115–1120. [CrossRef]
19. Ma, E.P.; Yiu, E.M. Multiparametric evaluation of dysphonic severity. *J. Voice* **2006**, *20*, 380–390. [CrossRef] [PubMed]
20. Siupsinskiene, N. Quantitative analysis of professionally trained versus untrained voices. *Medicina* **2003**, *39*, 36–46. [PubMed]
21. Leino, T.; Laukkanen, A.M.; Ilomaki, I.; Mäki, E. Assessment of vocal capacity of Finnish university students. *Folia Phoniatr. Logop.* **2008**, *60*, 199–209. [CrossRef] [PubMed]
22. D'Alatri, L.; Marchese, M.R. The speech range profile (SRP): An easy and useful tool to assess vocal limits. *Acta. Otorhinolaryngol. Ital.* **2014**, *34*, 253–258.
23. Dehqan, A.; Yadegari, F.; Scherer, R.C.; Dabirmoghadam, P. Correlation of VHI-30 to Acoustic Measurements across Three Common Voice Disorders. *J. Voice* **2017**, *31*, 34–40. [CrossRef]

International Journal of
Environmental Research and Public Health

MDPI

Article

Expression Profile of Stemness Markers CD138, Nestin and Alpha-SMA in Ameloblastic Tumours

Callisthenis Yiannis [1], Massimo Mascolo [2], Michele Davide Mignogna [3], Silvia Varricchio [2], Valentina Natella [2], Gaetano De Rosa [2], Roberto Lo Giudice [4], Cosimo Galletti [5], Rita Paolini [1] and Antonio Celentano [1,*]

1 Melbourne Dental School, University of Melbourne, 720 Swanston Street, Carton, VIC 3053, Australia; c.yiannis@student.unimelb.edu.au (C.Y.); rita.paolini@unimelb.edu.au (R.P.)
2 Pathology Unit, Department of Advanced Biomedical Sciences, University Federico II of Naples, 80131 Naples, Italy; massimo.mascolo@unina.it (M.M.); silvia.varricchio@gmail.com (S.V.); valenatella@gmail.com (V.N.); gaderosa@unina.it (G.D.R.)
3 Department of Neurosciences, Reproductive and Odontostomatological Sciences, University Federico II of Naples, 80131 Naples, Italy; mignogna@unina.it
4 Department of Clinical and Experimental Medicine, University of Messina, 98122 Messina, Italy; roberto.logiudice@unime.it
5 Comprehensive Dentistry Department, Faculty of Dentistry, Universitat de Barcelona, L'Hospitalet de Llobregat, 08007 Barcelona, Spain; cosimo88a@gmail.com
* Correspondence: antonio.celentano@unimelb.edu.au; Tel.: +61-393-411495

Abstract: Ameloblastic carcinoma is a rare malignant odontogenic neoplasm with a poor prognosis. It can arise de novo or from a pre-existing ameloblastoma. Research into stemness marker expression in ameloblastic tumours is lacking. This study aimed to explore the immunohistochemical expression of stemness markers nestin, CD138, and alpha-smooth muscle actin (alpha-SMA) for the characterisation of ameloblastic tumours. Six cases of ameloblastoma and four cases of ameloblastic carcinoma were assessed, including one case of ameloblastic carcinoma arising from desmoplastic ameloblastoma. In all tumour samples, CD138 was positive, whilst alpha-SMA was negative. Nestin was negative in all but one tumour sample. Conversely, the presence or absence of these markers varied in stroma samples. Nestin was observed in one ameloblastic carcinoma stroma sample, whilst CD138 was positive in one ameloblastoma case, one desmoplastic ameloblastoma case, and in two ameloblastic carcinoma stroma samples. Finally, alpha-SMA was found positive only in the desmoplastic ameloblastoma stroma sample. Our results suggest nestin expression to be an indicator for ameloblastic carcinoma, and CD138 and alpha-SMA to be promising biomarkers for the malignant transformation of ameloblastoma. Our data showed that nestin, CD138, and alpha-SMA are novel biomarkers for a better understanding of the origins and behaviour of ameloblastic tumours.

Keywords: ameloblastoma; ameloblastic carcinoma; nestin; CD138; syndecan-1; alpha-SMA; stemness markers

Citation: Yiannis, C.; Mascolo, M.; Mignogna, M.D.; Varricchio, S.; Natella, V.; De Rosa, G.; Lo Giudice, R.; Galletti, C.; Paolini, R.; Celentano, A. Expression Profile of Stemness Markers CD138, Nestin and Alpha-SMA in Ameloblastic Tumours. *Int. J. Environ. Res. Public Health* **2021**, *18*, 3899. https://doi.org/10.3390/ijerph18083899

Academic Editor: Takemi Otsuki

Received: 17 February 2021
Accepted: 6 April 2021
Published: 8 April 2021

1. Introduction

Ameloblastomas (ABs) are uncommon, benign, locally aggressive odontogenic tumours of epithelial origin with a high incidence of recurrence [1–4]. Left untreated, they have the potential to reach large sizes and cause physical disfiguration and functional disturbances. There is no gender preference, but there is a high incidence in the third and fourth decades of life [4,5]. Ameloblastic tumours show significant histological variations and are classified into various benign and malignant entities [1]. According to the WHO 2017 classification, benign ABs are categorised into conventional, unicystic, and extraosseous/peripheral types [6]. Conventional is the most common type and makes up 85% of cases [4,7]. Histologically, it can be categorised into follicular and plexiform [1,2]. Other less common histological variants are clear cell and desmoplastic cell [8]. Desmoplastic

ABs behave like conventional ABs, although their clinical and radiographic characteristics may be different [8–10].

In some cases, ABs can demonstrate metastasis with benign histological features. This type is classified as a metastasising (malignant) AB. This was originally classified as a malignant form in the 2005 WHO classification system, but has since been re-classified as a benign epithelial odontogenic tumour in the current 2017 WHO classification [6]. Ameloblastic carcinoma (AC) is an AB that can demonstrate metastatic histological features and malignant cytological characteristics. Accordingly, it is categorised as a malignant odontogenic tumour under the 2017 WHO classification [6]. Genetic and molecular alterations in these odontogenic epithelial tumours have been identified as possible associations with mechanisms of oncogenesis, cyto-differentiation, and tumour progression [11,12]. The development of an AC from an existing AB is extremely rare. Only 16 cases have been reported in literature in the last 10 years [13]. AC is generally characterised by high morbidity and mortality, and the survival rates of patients diagnosed with AC are significantly reduced in those with evidence of metastasis [14,15].

Stem cells are undifferentiated cells capable of self-renewal and the production of a diverse range of differentiated cells [16]. Tumours can contain a heterogenous population of stem cells known as cancer stem cells (CSCs). CSCs have the characteristics of self-renewal driving tumorigenesis [16,17]. Studies have suggested that haematopoietic and neural stem cell markers may play important roles in epithelial–mesenchymal interactions and cell proliferation/differentiation in both odontogenic epithelial tumours and during odontogenesis [18–21]. Specifically, several immunohistochemical studies have outlined an effective correlation between the levels of some markers of stemness such as nestin, CD138, and alpha-smooth muscle actin (alpha-SMA) and the different forms of ameloblastic tumours [22].

Nestin is an intermediate filament constituting the cytoskeleton, and is a marker of neural stem cells or progenitor cells [23]. The expression of nestin is related to tooth development and repair of dentine [24]. The expression of nestin in ameloblastomas and malignant ameloblastomas has been reported in the literature to be negative [24]. CD138, also known as syndecan-1, is a peptide that inhibits tumour growth. It is highly expressed in fibroblasts and epithelial cells [25]. Alterations in CD138 expression results in changes to cell behaviour, shape, growth, migration, and cytoskeletal organisation [26]. To date, CD138 expression and its role in AC remains debated. Alpha-SMA, a marker of myofibroblasts, has also been reported in studies comparing ABs to ACs [27]. A positive correlation has been identified between the number of myofibroblasts present in the stroma and the aggressive behaviour of odontogenic tumours through enhancement of epithelial–mesenchymal interactions [28,29]. Research has demonstrated that the clinical recurrence of ABs may be predicted by alpha-SMA expression [30]. CD138 and alpha-SMA expression may indicate a higher aggressive potential of AB [31], and alpha-SMA expression in epithelial cells may indicate AC [27,31,32].

Cases presenting with subtle metastatic change, atypical phenotypes, or poor biopsied tissue reduce the ability of an accurate diagnosis. The understanding of the role of stemness markers in ABs may aid in the early diagnosis of malignant ameloblastic tumours with direct implications on their management protocol [33].

Research into stemness marker expression in ameloblastic tumours is currently deficient and evidence is controversial. Further studies are required to better understand their role in this type of malignancy. The main aim of this study was to explore the immunoexpression of stem cell markers nestin, CD138 (syndecan-1), and alpha-SMA in a series of four cases of ameloblastic carcinoma and to compare this data to an ameloblastoma immunoexpression profile.

Our results suggest nestin expression to be an indicator of AC, and CD138 and alpha-SMA to be biomarkers for transformation of AC from AB. Our study confirms the role of nestin, CD138 (syndecan-1), and alpha-SMA as promising biomarkers for a better understanding of the origin and behaviour of ameloblastic tumours.

2. Materials and Methods

This study was approved by the Ethics Committee of the University "Federico II" of Naples, Italy (protocol n. 35/15). Appropriate permission and written informed consent were obtained from all the patients described in this article.

For this study, we selected a case of desmoplastic AB (female, 22 years of age) that progressed to AC, three further ACs (two males and a female; mean age: 57 years; age range: 48–73 years), and six ABs from the archives of the Pathology Section of Advanced Biomedical Sciences at the Federico II University of Naples, Italy; every patient had a follow-up of at least 24 months.

Each specimen was fixed in 10% buffered formalin, embedded in paraffin, and serially sectioned (4 μm thick sections). For each case, one section was stained with haematoxylin and eosin (H&E) and the others used for immunohistochemistry (labelled streptavidin-biotin standard technique) with anti-nestin (nestin, 10c2, Santa Cruz Biotechnology, Santa Cruz, CA, USA, diluted 1:100), anti-CD138/syndecan-1 (CD138, B-A38, Ventana, Tucson, AZ, USA, prediluted), and anti-α-smooth muscle actin (αSMA, 1A4, Ventana, Tucson, AZ, USA, prediluted) antibodies. Cells showing definite brown staining were judged positive for nestin, CD138, and alpha-SMA. All slides were examined in a double-blinded fashion by two pathologists (M.M. and G.D.R.) to confirm the diagnosis and to assess the immunohistochemical staining, both in tumour islands and stromal fibroblasts, according to a semiquantitative score, as negative, focal (+), moderate (++), and extensive (+++) positivity.

3. Results

Four cases of AC and six cases of AB were included in this study. Of these six, one case of desmoplastic AB progressed to AC. AC case descriptions can be found below.

Case 7: A 22-year-old female presented with a four-month history of an asymptomatic expansive lesion (bucco-lingually) involving quadrant 4, region 41–46 (Figure 1a–c). Significant loss of soft tissue and bony attachment (up to 100%) was evident. Increased vasculature and agenesis of the mandibular second premolars was observed (Figure 1a–c). The patient was a non-smoker and non-drinker. Diagnosis at this stage was a desmoplastic AB and treated with box resection including removal of 41–44 and 85. The one month follow-up indicated healthy and healing tissues (Figure 1d,e). The seven-month follow-up was characterised by the onset of an erythematous lesion on the buccal gingiva in the area of 31 (Figure 1f,g) and radiographic findings were also suggestive of a potential recurrence. Diagnosis at this stage was an AC and treated with a more extensive box resection including removal of 31 (Figure 1h). The histopathological description of this case is presented in Figure 2.

Case 8: A 48-year-old male presented with a history of a five-year lesion. Upon examination, the lesion was a 4 cm × 7 cm oral soft tissue ulceration with recurrent abscess and pus of the right mandible. Multiple teeth had been lost and swelling of the submandibular and lymph nodes of the neck were present homolaterally. The patient was a smoker (17 cigs/day) and a casual drinker. Treatment involved hemi-mandibulectomy. The histopathological description is presented in Figure 3a–d.

Case 9: A 73-year-old male presented with a solid tissue lesion involving destruction of the roof and anterior wall of the maxillary sinus infiltrating the nasal cavity and hard palate. The patient had a previous history (four years prior) of a basal cell carcinoma (BCC) and treatment involved excision of the entire right orbit. One year following treatment, the patient was diagnosed with a BCC of the right nose wing that was excised. The patient had hypertension and was a smoker (60 cigs/day). The histopathological description is presented in Figure 3e–h.

Case 10: A 50-year-old female presented with a carcinoma (33 mm in diameter) involving the right nasal fossa, ethmoidal bone, maxillary bone, maxillary sinus, and floor of the orbit. The patient was a non-smoker and non-drinker. The patient received maxillectomy and reconstruction with titanium mesh. The histopathological description is presented in Figure 3i–l.

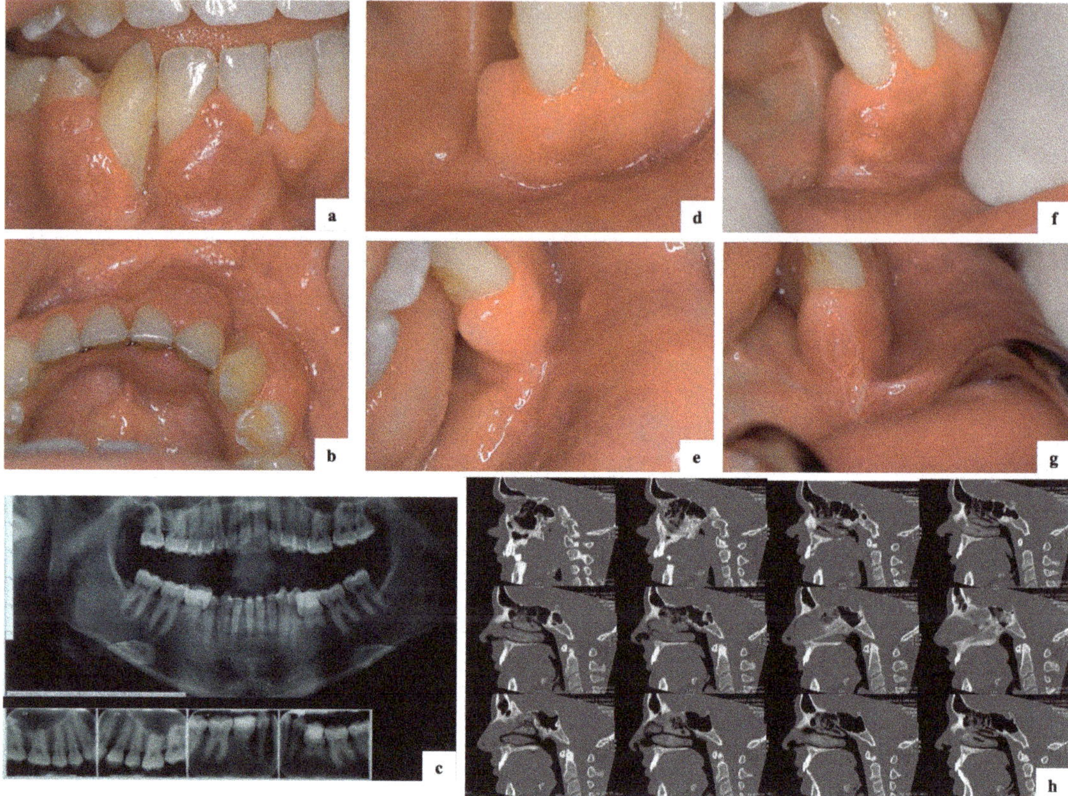

Figure 1. Initial presentation of a 22-year-old female with a desmoplastic ameloblastoma that progressed to an ameloblastic carcinoma (case 7). (**a–c**) Presentation with a four-month history of an asymptomatic expansive lesion (bucco-lingually) involving quadrant 4, region 41–46, with significant loss of soft tissue and bony attachment (up to 100%). Orthopantomogram (OPG) and peri-apical radiographs indicated agenesis of mandibular second premolars and retention of 75 and 85. Histopathologic diagnosis at this stage was a desmoplastic ameloblastoma. (**d,e**) Clinical presentation indicated healthy and healing tissues one-month after surgical intervention via box resection and removal of 41–44 and 85. (**f,g**) Clinical presentation at the seven-month follow-up indicated an erythematous lesion on the buccal gingiva in the area of 31. Diagnosis was an ameloblastic carcinoma and treated with a more extensive box resection including removal of 31. (**h**) Computed tomography (CT) scan one-month after surgery indicated removal of the ameloblastic carcinoma.

All histopathologic diagnoses of AB and AC were confirmed. ABs were typically composed of epithelial nests of columnar cells arranged in a palisading pattern surrounding a loose network of cells mimicking a stellate reticulum, in a loose connective stroma. ACs showed the typical features of malignancy, including marked nuclear atypia, a high mitotic index, and neural or vascular invasion, in the context of classical AB.

All ABs expressed CD138, with a variable degree of expression varying from + to +++, in their solid tumour counterpart; the stromal fibroblasts resulted negative, except for a case showing a moderate positivity (++) for CD138.

Nestin was negative in all AB cases. Alpha-SMA was negative in all AB cases except for one case (positive only in the stromal component).

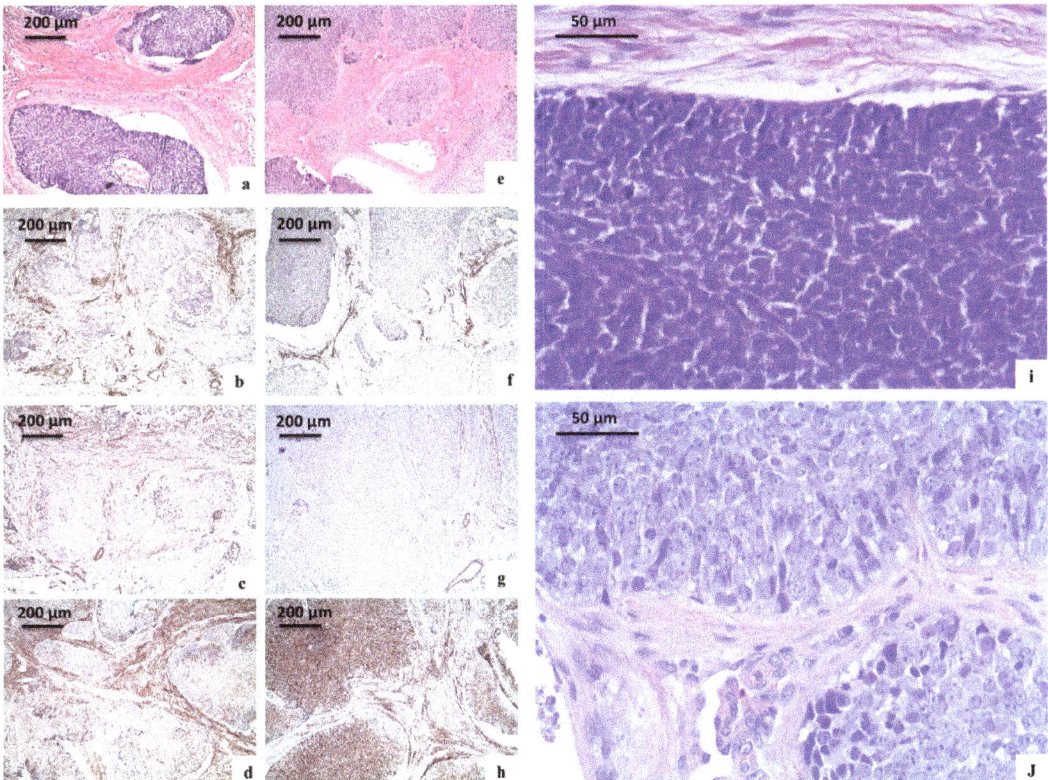

Figure 2. Immunohistochemical analysis of the stemness markers (nestin, CD138, and alpha-SMA) in an ameloblastic tumours. (**a–d**,**i**) Case 6; (**e–h**,**j**) case 7. (**a**) Follicular ameloblastoma (haematoxylin and eosin, original magnification ×100); (**b**) staining for nestin in follicular ameloblastoma (original magnification ×100); (**c**) staining for alpha-SMA in follicular ameloblastoma (original magnification ×100); (**d**) staining for CD138 in follicular ameloblastoma (original magnification ×100); (**e**) evolving desmoplastic ameloblastoma (haematoxylin and eosin, original magnification ×100); (**f**) staining for nestin in evolving desmoplastic ameloblastoma (original magnification ×100); (**g**) staining for alpha-SMA in evolving desmoplastic ameloblastoma (original magnification ×100); (**h**) staining for CD138 in evolving desmoplastic ameloblastoma (original magnification ×100); (**i**) follicular ameloblastoma: High magnification (haematoxylin and eosin, original magnification ×400); (**j**) evolving desmoplastic ameloblastoma: High magnification (haematoxylin and eosin, original magnification ×400).

The AC derived from the desmoplastic AB expressed alpha-SMA focally (+) in the smooth muscle around some tumour nests; CD138 diffusely stained the tumour islands (+++) and moderately (++) the stroma. Nestin was negative.

The three remaining AC cases were negative for alpha-SMA. Only one case showed an extensive (+++) positivity for nestin, both in the tumour cells and stromal fibroblasts.

Finally, one case of AC resulted positive for CD138, both in the tumour cells (++) and fibroblasts (+++), one case showed an extensive (+++) positivity only in the tumour cells, and one case was completely negative.

Table 1 summarises the immunostaining data.

Statistical analysis of these stemness markers did not show significant differences ($p < 0.05$; Fisher's exact test). This was expected due to the rarity of ACs and the consequent sample size in our cohort.

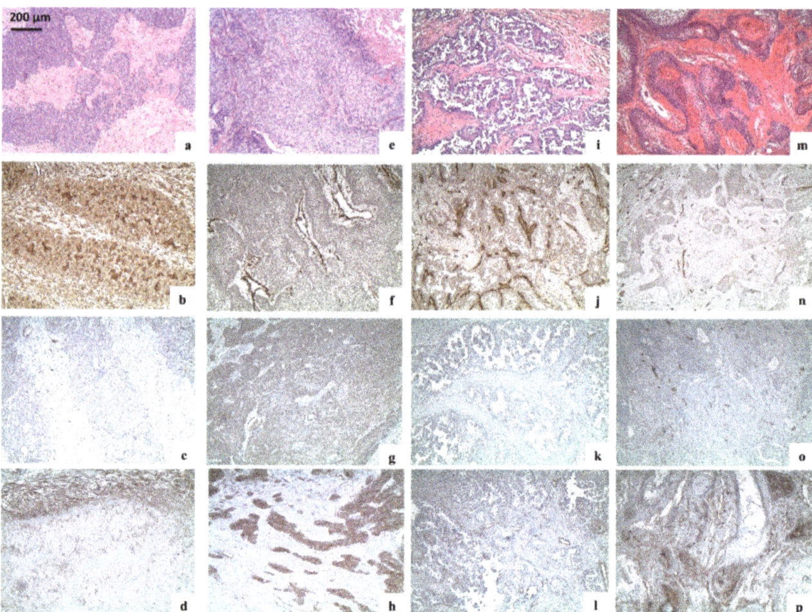

Figure 3. Immunohistochemical analysis of the stemness markers (nestin, CD138, and alpha-SMA) in ameloblastic tumours. (**a–d**) Case 8; (**e–h**) case 9; (**i–l**) case 10; (**m–p**) case 6. (**a**,**e**,**i**) Ameloblastic carcinomas (haematoxylin and eosin, original magnification ×100); (**m**) desmoplastic ameloblastoma (haematoxylin and eosin, original magnification ×100); (**b**,**f**,**j**) staining for nestin in ameloblastic carcinomas: Overexpression of nestin was seen in tumour cells, fibroblasts, endothelial cells, and lymphocytes (**b**: original magnification ×100); (**f**,**j**) tumour cells and fibroblasts were negative, while endothelial cells were positive (original magnification ×100); (**n**) staining for nestin in recurrent plexiform ameloblastoma: Tumour cells and fibroblasts were negative (original magnification ×100); (**c**,**g**,**k**) staining for alpha-SMA in ameloblastic carcinoma: Alpha-SMA was negative; a weak expression of alpha-SMA was present in the vessel wall (**c**,**g**,**k**: original magnification ×100); (**o**) staining for alpha-SMA in recurrent plexiform ameloblastoma: Tumour cells and fibroblasts were negative, but a weak expression was present in the vessel wall (original magnification ×100); (**d**,**h**,**l**) staining for CD138 in ameloblastic carcinoma: (**d**) Tumour cells and stromal fibroblasts stained an extensive and moderate expression of CD138, respectively (**d**: original magnification ×100); (**h**) overexpression of CD138 was seen only in tumour islands (**h**: original magnification ×100); (**l**) only a minority of tumour cells resulted positive for CD138 (**l**: original magnification ×100); (**p**) immunostaining for CD138 in recurrent plexiform ameloblastoma: Tumour islands were variably positive for CD138.

Table 1. Immunoexpression profile of all stemness markers (nestin, CD138, and alpha-smooth muscle actin (alpha-SMA)) in ameloblastic tumours.

Case n.	Diagnosis	Nestin		CD138 (Syndecan-1)		Alpha-SMA	
		Tumour	Stroma	Tumour	Stroma	Tumour	Stroma
1	Ameloblastoma	-	-	++	-	-	-
2	Ameloblastoma	-	-	+	-	-	-
3	Ameloblastoma	-	-	+++	-	-	-
4	Ameloblastoma	-	-	+	-	-	-
5	Ameloblastoma	-	-	++	-	-	-
6	Desmoplastic ameloblastoma	-	-	++	+++	-	++
7	Ameloblastic carcinoma derived from desmoplastic ameloblastoma	-	-	+++	++	-	+
8	Ameloblastic carcinoma	+++	+++	++	+++	-	-
9	Ameloblastic carcinoma	-	-	+++	-	-	-
10	Ameloblastic carcinoma	-	-	+	-	-	-

4. Discussion

ABs are uncommon, benign, locally aggressive odontogenic tumours of epithelial origin with a high incidence of recurrence [1–4] and potential for malignant transformation into a metastasising (malignant) AB or AC [15,34]. To date, there are approximately 65 cases of metastasising (malignant) AB and 125 cases of AC reported in the literature [15,34]. ABs present radiographically as uni- or multilocular radiolucencies frequently with cortical expansion. Clinicians should suspect a malignant lesion when there are presentations of paraesthesia, pain, irregular borders, and invasion of adjacent tissues [14,35].

ACs usually present with microscopic evidence of malignancy. However, confirmation of whether this is a secondary AC must be acknowledged by a history of persistent, recurrent or residual AB [36]. Differential diagnosis between the lesions, from incisional biopsies, can prove to be difficult. Differential diagnosis often includes other types of intra-osseous carcinomas of the jaws [36]. Immunohistochemistry can be utilised to aid in the diagnosis and classification of odontogenic tumours. Some currently published immunohistochemical markers used to differentiate AC from AB include CK18, parenchymal MMP-2, stromal MMP-9, Ki-67, and p53 [36,37]. Novel immunomarkers are crucial for a better understanding of lesion origins, diagnosis, and behaviour.

Diagnostic difficulty is sometimes encountered with ABs of unusually aggressive behaviour and in differentiation from ACs. To identify the markers of stemness that will have implications in the diagnosis of AC and cases with subtle malignant transformation, immunohistochemical expression of a panel of markers (nestin, CD138, and alpha-SMA) was investigated.

Nestin is a CSC surface marker identified in tissues and pathological conditions, such as the neural crest, heart, testis, reactive astrocytes (after brain injury), and the central and peripheral nervous systems [24]. Previous studies have discussed the expression of nestin as a useful marker for the identification of odontogenic ectomesenchyme and odontoblasts in odontogenic tumours [24,38]. Previously published data by Fujita et al. (2006) indicated that almost all of their cases of ameloblastomas and malignant ameloblastomas (three cases) were negative for nestin [24]. This was reflected in our experimental study. However, amongst our cases of AC, 25% of our cohort had an extensive expression of nestin in both the tumour islands and stromal cells. Analogous to the literature, there was no expression of nestin in any of our AB cases. Although our sample size was small and further studies should aim to better elicit its expression in larger cohorts, nestin appears to have immunoreactivity in malignant ameloblastic tumours.

CD138 is highly expressed in fibroblasts and epithelial cells and functions to inhibit tumour growth [38]. The expression of CD138 in ABs and ACs is controversial amongst the literature. Various studies have reported different findings in regard to increased or decreased expression in ACs. A study published by Bologna-Molina et al. (2009) suggested CD138 expression in desmoplastic AC to be inversely correlated to the proliferative index Ki67 [39]. Therefore, according to these authors, decreased CD138 expression in desmoplastic AB corresponds with its higher aggressiveness [39]. However, a study published by Martínez-Martínez et al. (2017) indicated CD138 is mainly expressed in the peripheral cells of ABs and is expressed in most areas of AC [36]. The AB cases in our cohort indicated the expression of CD138 (from focal to extensive) in the solid tumour but negative in the stellate tumour areas and stromal fibroblasts. The case of desmoplastic AB resulted positive in fibroblasts and tumour islands. A particular focus of our study was to assess the progression of AC from a desmoplastic AB. The AC arising from the desmoplastic AB showed CD138 to have greater positivity in the tumour islands but less positivity in the stroma. All cases of AC were positive for CD138 in the tumour, and 50% of cases were positive in the stroma. Our data suggest that ameloblastic tumours may be positive for CD138 in tumour islands. This is in accordance with research published by Martínez-Martínez [36]. Furthermore, developing AC from desmoplastic ABs may be positive for CD138 in tumour islands and stromal cells. Further studies should aim to confirm these results as a biomarker for transformation of AC from AB.

Alpha-SMA, a marker of myofibroblasts, has been assessed in the literature for its potential to determine the recurrence of AB and as a biomarker of transformation to AC [27,30–32]. Interestingly, in our experimental study, only the case of desmoplastic AB and the developing AC resulted positive for alpha-SMA. This was only in the surrounding tumour nests. All other cases of AB and AC were negative for alpha-SMA. A study published by Siar and Ng (2019) assessed the epithelial–mesenchymal transition of neoplastic cells, as it is essential for metastatic expansion and cancer progression [40]. It was concluded that stromal upregulation of alpha-SMA (as well as osteonectin and N-cadherin) implicates a role in local invasiveness [40]. Further research suggests a positive correlation between the number of myofibroblasts present in the stroma and the aggressive behaviour of odontogenic tumours [28]. This suggests a histopathological feature for a developing AC. Considering alpha-SMA expression in epithelial cells may indicate AC [27,31,32,41], further studies should aim to assess alpha-SMA expression in fibroblasts and epithelial cells.

5. Conclusions

The present study had the opportunity to assess the rare progression of AC from a desmoplastic AB. The immunohistochemical results suggested nestin expression to be an indicator for AC, and CD138 and alpha-SMA to be biomarkers for transformation of AC from AB. However, further experiments are required to look at tumour and stroma tissue in order to elucidate the presence or absence of these biomarkers. This will allow a better understanding of the expression of stemness markers amongst a larger cohort of cases and progressively shed light on their immunoexpression amongst ameloblastic tumours.

Author Contributions: Conceptualisation, A.C.; data curation, C.Y., M.M., and A.C.; formal analysis, M.M., G.D.R., and A.C.; investigation, C.Y., M.M., S.V., V.N., and A.C.; methodology, M.M., M.D.M., and G.D.R.; project administration, A.C.; resources, M.M., G.D.R., and A.C.; software, M.D.M. and A.C.; supervision, M.M., M.D.M., and A.C.; validation, M.M., G.D.R., and A.C.; visualisation, S.V., V.N., and A.C.; writing—original draft, C.Y., M.M., R.P., and A.C.; writing—review and editing, C.Y., R.L.G., C.G., R.P., and A.C. All authors have read and agreed to the published version of the manuscript.

Funding: This research received no external funding.

Institutional Review Board Statement: This study was approved by the Ethics Committee of University "Federico II" of Naples (protocol n. 35/15).

Informed Consent Statement: Informed consent was obtained from all subjects involved in the study.

Data Availability Statement: The data presented in this study are available on request from the corresponding author.

Acknowledgments: The authors gratefully acknowledge the support of the Melbourne Dental School, The University of Melbourne, and of the University "Federico II" of Naples, Naples, Italy.

Conflicts of Interest: The authors declare no conflict of interest.

References

1. Philipsen, H.P.; Reichart, P.A.; Slootweg, P.J. Odontogenic tumours. In *WHO Classification of Tumours. Pathology and Genetics. Head and Neck Tumours*; Barnes, L., Eveson, J.W., Reichart, P.A., Sidransky, D., Eds.; IARC Press: Lyon, France, 2005; pp. 283–327.
2. Sciubba, J.J. *Tumors and Cysts of the Jaw*; American Registry of Pathology: Silver Spring, MD, USA, 2001; pp. 19–22.
3. Milman, T.; Ying, G.S.; Pan, W.; Li Volsi, V. Ameloblastoma: 25 Year Experience at a Single Institution. *Head Neck Pathol.* **2016**, *10*, 513–520. [CrossRef] [PubMed]
4. Sham, E.; Leong, J.; Maher, R.; Schenberg, M.; Leung, M.; Mansour, A.K. Mandibular ameloblastoma: Clinical experience and literature review. *ANZ J. Surg.* **2009**, *79*, 739–744. [CrossRef] [PubMed]
5. Bassey, G.O.; Osunde, O.D.; Anyanechi, C.E. Maxillofacial tumors and tumor-like lesions in a Nigerian teaching hospital: An eleven year retrospective analysis. *Afr. Health Sci.* **2014**, *14*, 56–63. [CrossRef]
6. El-Naggar, A.K.; Chan, J.K.C.; Grandis, J.R.; Takata, T.; Slootweg, P.J. *WHO Classification of Head and Neck Tumours*, 4th ed.; International Agency for Research on Cancer (IARC): Lyon, France, 2017; Volume 9.
7. Hertog, D.; Bloemena, E.; Aartman, I.H.; van-der-Waal, I. Histopathology of ameloblastoma of the jaws; some critical observations based on a 40 years single institution experience. *Med. Oral Patol. Oral Cir. Bucal* **2012**, *17*, e76–e82. [CrossRef]

8. Cadavid, A.M.H.; Araujo, J.P.; Coutinho-Camillo, C.M.; Bologna, S.; Junior, C.A.L.; Lourenço, S.V. Ameloblastomas: Current aspects of the new WHO classification in an analysis of 136 cases. *Surg. Exp. Pathol.* **2019**, *2*, 17. [CrossRef]
9. Speight, P.M.; Takata, T. New tumour entities in the 4th edition of the World Health Organization Classification of Head and Neck tumours: Odontogenic and maxillofacial bone tumours. *Virchows Arch.* **2018**, *472*, 331–339. [CrossRef] [PubMed]
10. Dias, C.D.; Brandão, T.B.; Soares, F.A.; Lourenço, S.V. Ameloblastomas: Clinical-histopathological evaluation of 85 cases with emphasis on squamous metaplasia and keratinization aspects. *Acta Odontol. Scand.* **2013**, *71*, 1651–1655. [CrossRef]
11. Heikinheimo, K.; Jee, K.J.; Niini, T.; Aalto, Y.; Happonen, R.P.; Leivo, I.; Knuutila, S. Gene expression profiling of ameloblastoma and human tooth germ by means of a cDNA microarray. *J. Dent. Res.* **2002**, *81*, 525–530. [CrossRef]
12. Kumamoto, H. Molecular pathology of odontogenic tumors. *J. Oral Pathol. Med.* **2006**, *35*, 65–74. [CrossRef]
13. Costa Neto, H.; Carmo, A.F.; Andrade, A.L.D.L.; Rodrigues, R.R.; Germano, A.R.; Freitas, R.A.; Galvão, H.C. Ameloblastic carcinoma arising from a preexistent ameloblastoma. *J. Bras. Patol. Med. Lab.* **2019**, *55*, 530–539. [CrossRef]
14. Yoon, H.J.; Hong, S.P.; Lee, J.I.; Lee, S.S.; Hong, S.D. Ameloblastic carcinoma: An analysis of 6 cases with review of the literature. *Oral Surg. Oral Med. Oral Pathol. Oral Radiol. Endod.* **2009**, *108*, 904–913. [CrossRef]
15. Deng, L.; Wang, R.; Yang, M.; Li, W.; Zou, L. Ameloblastic carcinoma: Clinicopathological analysis of 18 cases and a systematic review. *Head Neck* **2019**, *41*, 4191–4198. [CrossRef] [PubMed]
16. Reya, T.; Morrison, S.J.; Clarke, M.F.; Weissman, I.L. Stem cells, cancer, and cancer stem cells. *Nature* **2001**, *414*, 105–111. [CrossRef]
17. Jordan, C.T.; Guzman, M.L.; Noble, M. Cancer stem cells. *N. Engl. J. Med.* **2006**, *355*, 1253–1261. [CrossRef]
18. Kumamoto, H.; Ohki, K. Detection of Notch signaling molecules in ameloblastomas. *J. Oral Pathol. Med.* **2008**, *37*, 228–234. [CrossRef] [PubMed]
19. Kumamoto, H.; Ohki, K.; Ooya, K. Expression of Sonic hedgehog (SHH) signaling molecules in ameloblastomas. *J. Oral Pathol. Med.* **2004**, *33*, 185–190. [CrossRef] [PubMed]
20. Kumamoto, H.; Ooya, K. Immunohistochemical detection of beta-catenin and adenomatous polyposis coli in ameloblastomas. *J. Oral Pathol. Med.* **2005**, *34*, 401–406. [CrossRef] [PubMed]
21. Kumamoto, H.; Ooya, K. Expression of bone morphogenetic proteins and their associated molecules in ameloblastomas and adenomatoid odontogenic tumors. *Oral Dis.* **2006**, *12*, 163–170. [CrossRef]
22. Kallianpur, S.; Jadwani, S.; Misra, B.; Sudheendra, U.S. Ameloblastic carcinoma of the mandible: Report of a case and review. *J. Oral Maxillofac. Pathol.* **2014**, *18*, S96–S102. [CrossRef]
23. Ando, T.; Shrestha, M.; Nakamoto, T.; Uchisako, K.; Yamasaki, S.; Koizumi, K.; Ogawa, I.; Miyauchi, M.; Takata, T. A case of primordial odontogenic tumor: A new entity in the latest WHO classification (2017). *Pathol. Int.* **2017**, *67*, 365–369. [CrossRef]
24. Fujita, S.; Hideshima, K.; Ikeda, T. Nestin expression in odontoblasts and odontogenic ectomesenchymal tissue of odontogenic tumours. *J. Clin. Pathol.* **2006**, *59*, 240–245. [CrossRef]
25. Sanderson, R.D.; Hinkes, M.T.; Bernfield, M. Syndecan-1, a cell-surface proteoglycan, changes in size and abundance when keratinocytes stratify. *J. Investig. Dermatol.* **1992**, *99*, 390–396. [CrossRef] [PubMed]
26. Wijdenes, J.; Dore, J.M.; Clement, C.; Vermot-Desroches, C. CD138. *J. Biol. Regul. Homeost. Agents* **2002**, *16*, 152–155.
27. Kamath, K.P.; Vidya, M.; Shetty, N.; Karkera, B.V.; Jogi, H. Nucleolar organizing regions and alpha-smooth muscle actin expression in a case of ameloblastic carcinoma. *Head Neck Pathol.* **2010**, *4*, 157–162. [CrossRef] [PubMed]
28. Vered, M.; Shohat, I.; Buchner, A.; Dayan, D. Myofibroblasts in stroma of odontogenic cysts and tumors can contribute to variations in the biological behavior of lesions. *Oral Oncol.* **2005**, *41*, 1028–1033. [CrossRef] [PubMed]
29. De Wever, O.; Mareel, M. Role of myofibroblasts at the invasion front. *Biol. Chem.* **2002**, *383*, 55–67. [CrossRef]
30. Mahmoud, S.A.M.; Amer, H.W.; Mohamed, S.I. Primary ameloblastic carcinoma: Literature review with case series. *Pol. J. Pathol.* **2018**, *69*, 243–253. [CrossRef]
31. Safadi, R.A.; Quda, B.F.; Hammad, H.M. Immunohistochemical expression of K6, K8, K16, K17, K19, maspin, syndecan-1 (CD138), α-SMA, and Ki-67 in ameloblastoma and ameloblastic carcinoma: Diagnostic and prognostic correlations. *Oral Surg. Oral Med. Oral Pathol. Oral Radiol.* **2016**, *121*, 402–411. [CrossRef]
32. Bello, I.O.; Alanen, K.; Slootweg, P.J.; Salo, T. Alpha-smooth muscle actin within epithelial islands is predictive of ameloblastic carcinoma. *Oral Oncol.* **2009**, *45*, 760–765. [CrossRef]
33. Khan, W.; Augustine, D.; Rao, R.S.; Sowmya, S.V.; Haragannavar, V.C.; Nambiar, S. Stem Cell Markers SOX-2 and OCT-4 Enable to Resolve the Diagnostic Dilemma between Ameloblastic Carcinoma and Aggressive Solid Multicystic Ameloblastoma. *Adv. Biomed. Res.* **2018**, *7*, 149. [CrossRef]
34. Hosalkar, R.; Saluja, T.S.; Swain, N.; Singh, S.K. Prognostic evaluation of metastasizing ameloblastoma: A systematic review of reported cases in literature. *J. Stomatol. Oral Maxillofac. Surg.* **2020**. [CrossRef]
35. Fregnani, E.R.; da Cruz Perez, D.E.; de Almeida, O.P.; Kowalski, L.P.; Soares, F.A.; de Abreu Alves, F. Clinicopathological study and treatment outcomes of 121 cases of ameloblastomas. *Int. J. Oral Maxillofac. Surg.* **2010**, *39*, 145–149. [CrossRef]
36. Martínez-Martínez, M.; Mosqueda-Taylor, A.; Carlos-Bregni, R.; Pires, F.R.; Delgado-Azanero, W.; Neves-Silva, R.; Aldape-Barrios, B.; Paes-de Almeida, O. Comparative histological and immunohistochemical study of ameloblastomas and ameloblastic carcinomas. *Med. Oral Patol. Oral Cir. Bucal* **2017**, *22*, e324–e332. [CrossRef] [PubMed]
37. Yoon, H.J.; Jo, B.C.; Shin, W.J.; Cho, Y.A.; Lee, J.I.; Hong, S.P.; Hong, S.D. Comparative immunohistochemical study of ameloblastoma and ameloblastic carcinoma. *Oral Surg. Oral Med. Oral Pathol. Oral Radiol. Endod.* **2011**, *112*, 767–776. [CrossRef]
38. Premalatha, B.R.; Patil, S.; Rao, R.S.; Reddy, N.P.; Indu, M. Odontogenic tumor markers—An overview. *J. Int. Oral Health* **2013**, *5*, 59–69.

39. Bologna-Molina, R.; Mosqueda-Taylor, A.; Lopez-Corella, E.; de Almeida, O.P.; Carrasco-Daza, D.; Farfán-Morales, J.E.; Molina-Frechero, N.; Damián-Matsumura, P. Comparative expression of syndecan-1 and Ki-67 in peripheral and desmoplastic ameloblastomas and ameloblastic carcinoma. *Pathol. Int.* **2009**, *59*, 229–233. [CrossRef]
40. Siar, C.H.; Ng, K.H. Epithelial-to-mesenchymal transition in ameloblastoma: Focus on morphologically evident mesenchymal phenotypic transition. *Pathology* **2019**, *51*, 494–501. [CrossRef] [PubMed]
41. Roy, S.; Garg, V. Alpha smooth muscle actin expression in a case of ameloblastic carcinoma: A case report. *J. Oral Maxillofac. Res.* **2013**, *4*, e4. [CrossRef] [PubMed]

International Journal of
Environmental Research and Public Health

MDPI

Article

The Formation of Biofilm and Bacteriology in Otitis Media with Effusion in Children: A Prospective Cross-Sectional Study

Artur Niedzielski [1,2,*], Lechosław Paweł Chmielik [1] and Tomasz Stankiewicz [2]

[1] Clinic of Pediatric Otolaryngology, Center of Postgraduate Medical Education (CMKP), 01-813 Warsaw, Poland; l.p.chmielik@chmielik.pl

[2] Independent Otoneurological Laboratory, Medical University of Lublin, 20-093 Lublin, Poland; tomaszstankiewicz@umlub.pl

* Correspondence: arturniedzielski@umlub.pl; Tel.: +48-225690235; Fax: +48-228341820

Abstract: Background: Otitis media with effusion (OME) can cause serious complications such as hearing impairment or development delays. The aim of the study was to assess the microbiological profile of organisms responsible for OME and to determine if a biofilm formation can be observed. Methods: Ninety-nine samples from 76 patients aged from 6 months to 12 years were collected for microbiological and molecular studies. Results: In microbiological studies, pathogenic bacteria *Haemophilus influenzae* (38.89%), *Streptococcus pneumoniae* (33.33%), and *Staphylococcus aureus MSSA* (27.78%), as well as opportunistic bacteria *Staphylococcus* spp. (74.14%), *Diphtheroids* (20.69%), *Streptococcus viridans* (3.45%), and *Neisseria* spp. (1.72%) were found. The average degree of hearing loss in the group of children with positive bacterial culture was 35.9 dB, while in the group with negative bacterial culture it was 25.9 dB ($p = 0.0008$). The type of cultured bacteria had a significant impact on the degree of hearing impairment in children ($p = 0.0192$). In total, 37.5% of *Staphylococcus* spp. strains were able to form biofilm. Conclusions: *Staphylococcus* spp. in OME may form biofilms, which can explain the chronic character of the disease. Pathogenic and opportunistic bacteria may be involved in the etiopathogenesis of OME. The degree of hearing loss was significantly higher in patients from which the positive bacterial cultures were obtained.

Keywords: otitis media with effusion; child; biofilms

Citation: Niedzielski, A.; Chmielik, L.P.; Stankiewicz, T. The Formation of Biofilm and Bacteriology in Otitis Media with Effusion in Children: A Prospective Cross-Sectional Study. *Int. J. Environ. Res. Public Health* **2021**, *18*, 3555. https://doi.org/10.3390/ijerph18073555

Academic Editors: Francesco Galletti, Francesco Gazia, Francesco Freni and Cosimo Galletti

Received: 1 March 2021
Accepted: 24 March 2021
Published: 30 March 2021

1. Introduction

Otitis media with effusion (OME) is a chronic inflammatory condition of the middle ear without general symptoms of acute infection. The disease is characterized by the presence of fluid in the tympanic cavity and conductive hearing loss. OME is one of the most common diseases in childhood. Two-thirds of children have had at least one episode of OME by the age of 3 years. One-third of them will have the attack without notice; therefore, it is called "silent" otitis media and can impair their hearing secretly [1]. It is also the most common cause of hearing loss in the pediatric population, which may adversely affect the development of speech as well as linguistic and cognitive abilities [2].

The pathogenesis of the disease is not fully understood and is most likely multifactorial. The development of recurrent and chronic ear infections is influenced by individual and environmental factors [3]. The individual risk factors for exudative otitis include: (1) anatomical and functional dysfunction of the eustachian tube [4]; (2) genetic predisposition [5]; (3) male gender [6]; (4) recurrent infections of the upper respiratory tract [7]; (5) episode of acute otitis media in the first 6 months of life [8]; (6) developmental defects in the craniofacial region, especially cleft palate and abnormal structure of the mastoid process with impaired pneumatization [9,10]; (7) the overgrowth of the Waldeyer's tonsillar ring [11]; (8) gastroesophageal reflux disease [12]; (9) immunological disorders [13]; and (10) allergy [14]. The factors of increased risk of OME in relation to environmental factors [15] play the most important role: (1) exposure to tobacco smoke; (2) attendance at a day care center (nursery, kindergarten); (3) poor socioeconomic status; and (4) autumn and winter seasons [15].

In most exudates in acute otitis media, *Haemophilus influenzae, Streptococcus pneumoniae*, and *Moraxella catarrhalis* are found. On the other hand, cultures of exudate in OME indicate no bacteria. Moreover, exudate is resistant to antibiotic treatment and susceptible to many inflammatory mediators, which led to the concept of its sterility. However, polymerase chain reaction (PCR) improved the sensitivity of bacterial detection in middle ear infections and is useful for the detection of pathogens that are slowly growing, difficult to culture, or hazardous to handle in a diagnostic lab [16]. Reason for a better detection of bacteria with PCR may be due to a small number of microorganisms that do not reach the limit of detection by direct culturing [17–26].

Using PCR technique, in 20–50% of the middle ear exudates, the *Alloiococcus otitidis* bacterium was detected. It has been found that this bacterium has a high immunostimulatory capacity and promotes colonization of the middle ear space [27,28].

It has long been known that 70% of the cultures are sterile in OME. Numerous reports indicate the lack of effects of antibiotic therapy in OME, indicating biofilm as the causative agent of the chronic nature of the disease [29].

Bacteria in nature often exist as sessile communities called biofilms [30]. These communities develop structures that are morphologically and physiologically differentiated from free-living bacteria [30] and more resistant to external factors. The concept of disease based on the existence of bacterial biofilm [31–35] explains why in a chronic bacterial infection it may be difficult to obtain positive cultures and explains the relative failure of treatment of OME with antibacterial drugs [30]. However pharmacokinetic drug penetration studies consistently indicate that the bactericidal concentration of the drug is easily achieved in the ear, and studies on planktonic bacteria indicate their sensitivity in vitro to antibacterial agents. The mucosal biofilm hypothesis also explains the observation that the most effective treatment of OME is tympanostomy with ear drainage. The poorly ventilated middle ear is an excellent environment for the formation of a bacterial mucosal biofilm. Ventilation tube surgery is an effective procedure for the following reasons: (1) restoration of middle ear ventilation increases oxygen concentration in this area, potentially changing the biofilm phenotype, (2) mechanical suction of exudate after tympanic membrane cutting interrupts continuity, cleanses, and reduces the mass of biofilm, and (3) restoration of ventilation facilitates reconstructing the host's defensive mechanisms in the mucosa of the middle ear. These changes lead to purification of the biofilm and remission of exudate [21,36,37].

The aim of the work was to evaluate the microorganisms responsible for otitis media with effusion development and their ability to form biofilms.

2. Materials and Methods

Our study included 76 patients admitted for surgical treatment of otitis media with effusion. Among 76 children participating in the study, 44 were boys (57.9%) and 32 were girls (42.1%), and their age ranged from 6 months to 12 years. All patients underwent basic laryngological and hearing evaluation according to age (impedance audiometry, otoacoustic emissions, pure-tone threshold audiometry). Impedance audiometry [38] was performed with the Madsen Zodiak 901 clinical tympanometer (GN Otometrics, Taastrup, Denmark), and otoacoustic emissions [39] were tested by the OtoRead™ Otoacoustic Emission Test Instrument (Interacoustics, Middelfart, Denmark). Air and bone conduction pure-tone auditory [40] threshold measurement was performed in cooperating patients using the Madsen Orbiter 922-2 clinical audiometer (GN Otometrics, Denmark). Based on the hearing evaluation, patients underwent either unilateral or bilateral tympanotomy. During the surgery we collected 99 samples from the middle ear space. From patients with bilateral OME, due to the small amount of the exudate, we collected two samples from the same patient in order to perform all needed examinations. Each sample was divided into two parts, one of which was used for microbiological culture and the other for molecular biology.

After cleaning and disinfection with a 70% spirit solution of the external auditory canal and the eardrum, tympanic membrane was cut in the posterior-lower quadrant using an operating microscope. The material was collected under pressure using a Polymed Mucus Extractor disposable sterile collection set for Poly Pedicure Ltd. (Ballabhgarh, India).

Middle ear aspirates were inoculated onto Columbia medium with 5% sheep blood, Chapman medium, MacConkey agar, chocolate agar with bacitracin, and Sabouraud agar. Then media were incubated at 37 °C for 24 h under aerobic conditions. Only chocolate agar with bacitracin was incubated under microaerophilic conditions 5–10% CO_2 using an anaerostat, and Sabouaraud agar was incubated at 30 °C for up to 5 days. After incubation, macroscopic assessment of the growth of bacteria and fungi on the media was made and colony morphology was determined (i.e., shape, size, surface, color, transparency).

To identify species of the bacteria of the Staphylococcus genus catalase test [41], the Slidex Staph-Kit agglutination test (bioMerieux, Marcy-l'Étoile, France) [42], coagulase free test [43], and the API Staph system tests (bioMerieux) [44] were performed.

To detect bacteria of the *Streptococcus* genus, the following tests we performed: micro-screen Strep latex confirmation assay (Lab M, Neogen, Heywood, UK), optochin resistance test [45], and the API 20 Strep system tests (bioMerieux) [46].

Identification of *Haemophilus* species was made using X and V growth factor requirement tests. Depending on the species, bacteria need separate growth factors X (hemin or hematine) and V (NAD or NADP) to develop. Firstly, the plates were brought to room temperature and then the pure culture of the *Haemophilus* strain was suspended in a sterile 0.9% NaCl solution to obtain a suspension with a density of about 0.5 McFarland. After the suspension was inoculated with a sterile cotton swab onto Muller–Hinton agar standardized according to National Committee for Clinical Laboratory Standards (NCCLS) recommendations, the diagnostic discs BVX, BV, and BX (B-bacitracin, factor V, factor X) were plated at a distance of about 15–20 mm apart from each other. Then the material was incubated at 35 °C for 18–24 h in an atmosphere of 5–7% CO_2.

Molecular microbial analysis was performed based on the polymerase chain reaction (PCR), which consists of the following stages: isolation of genomic DNA from cells present in the exudate and from coagulase positive staphylococci grown from these fluids, amplification of a specific fragment of the isolated genetic material of the microorganism, and detection of the amplified product. Genomic DNA was isolated from exudate and *Staphylococcus sp.* using the manufacturers protocol. The amount and purity of the DNA was checked using an Eppendorf BioPhotometer (Eppendorf AG, Hamburg, Germany). The largest amount of genomic DNA obtainable using the test described above was 60 µg. The purity of the sample was determined as the ratio of absorbance at 260 nm and 280 nm and for pure DNA (the index A260/280 is: 1.7–1.9). Measurements were made in disposable cuvettes with the Eppendorf BioPhotometer measurement instructions.

For samples with negative culture results ($n = 61$), a PCR reaction was performed. The DNA fragment encoding the bacterial 16s rRNA subunit in the nested-PCR system was amplified. The test was based on two amplification reactions: outer (Table 1), in which the resulting product was 740 base pairs; and a nested reaction (Table 2), in which the resulting product was 290 base pairs.

Table 1. Thermal profile of the first, outer amplification reaction (40 cycles from 2 to 4).

Type of Reaction	Time	Temperature
initial denaturation	2 min	95 °C
denaturation	30 s	95 °C
annealing step	60 s	58 °C
elongation	45 s	72 °C
final elongation	6 min	72 °C

Table 2. Thermal profile of the second, nested amplification reaction (40 cycles from 2 to 4).

Type of Reaction	Time	Temperature
initial denaturation	1 min	95 °C
denaturation	30 s	95 °C
annealing step	30 s	50 °C
elongation	30 s	72 °C
final elongation	6 min	72 °C

The resulting PCR product of 740 base pairs could or could not be detected on an agarose gel stained with ethidium bromide. This PCR product, however, was a template for the nested reaction in which primers complementary to the sites within the 740 bp product were used. Thus, a product of 290 base pairs was formed, which could be visible in the agarose gel in the form of a clear band for a positive sample. The nested reaction was carried out in the case of a negative result of the outer reaction.

The composition of the outer PCR reaction mixture for one sample included: 39.0 µL master mix PCR-out, 5.0 µL dNTPs mixture, 5.0 µL DNA, and 1.0 µL of Delta2 polymerase. The nested PCR reaction mixture for one sample consisted of: 39.0 µL master PCR-In, 5.0 µL dNTPs mixture, 5.0 µL of the PCR product obtained in the initial amplification, and 1.0 µL of Delta2 polymerase. Primers used in outer and nested PCR reactions were described by Gok et al. [47]. The amplified product was detected with 2% agarose gel electrophoresis [48]. In our study we used 10 µL of amplification product and 3 µL of dye (bromophenol blue). To determine the position of the amplification reaction product, a DNA size marker was used: ΦX174 DNA/BsuRI (MBI, Fermentas, Lithuania). For each PCR reaction positive and negative controls were performed.

The ability to form biofilm by coagulase-negative staphylococci (CNS) strains was tested by using Congo Red Agar method (CRA), Tissue Culture Plate (TCP) and determination of the presence of the *ica* operon genes in CNS strains.

Congo Red Agar method (CRA), prepared as described by Freemen and colleagues in 1989, was used to determine if coagulase-negative staphylococci can form biofilm. Firstly, plates with medium were inoculated and incubated to obtain single colonies. After 24 h at 37 °C, positive strains appeared as black colonies with a dry, crystalline consistency, while the polysaccharide negative strains remained red.

Tissue Culture Plate (TCP) is considered the gold standard phenotypic method of biofilm detection. In this method, bacterial adherence is measured spectrophotometrically [49]. In our study, a suspension of bacteria with a density of 0.5 MF (Mc Farland) was incubated at 37 °C for 24 h under aerobic conditions, then diluted in 1: 100 TSB buffer. A portion of 100 µL of the suspension of each strain (3 replicates for one strain) was applied onto the microplate and incubated at 37 °C for 24 h in aerobic conditions. The plate was rinsed 3 times with TSB buffer and then the plate was stained with 0.1% crystal violet for 15 min. After washing the dye, 100 µL of absolute alcohol was added to each well. The absorbance reading was performed in an ELISA reader at 570 nm. As a negative control, we used a *Staphylococcus epidermidis* ATCC 12228 reference strain, which does not produce a biofilm. The positive result was the absorbance greater than twice the mean absorbance value read for the negative control.

Determination of the presence of the intercellular adhesion (*ica*) operon genes in cultured coagulase-resistant staphylococci strains was made as follows. Genomic DNA was isolated from exudate and *Staphylococcus sp.* using the manufacturer's protocol. In order to identify the presence of *ica* operon genes in coagulase-negative staphylococci, PCR reactions were performed using primers for the *icaA*, *icaB*, *icaC*, and *icaD* genes. The primer sequences for each of the *ica* operon genes, the amplification conditions for each pair of primers, and the size of the amplification products are shown in the following studies of Ziebuhr [50] and de Silva [51] (Table 3).

Table 3. Primer sequences, amplification conditions, and sizes of amplification products of *icaABCD* genes for *Staphylococcus epidermidis*.

PCR Product	Primer Sequences	Amplification Conditions	Product Size
icaA	Forward: 5′-GACCTCGAAGTC AATAGAGGT Reverse: 5′-CCCAGTATAACGTTGGATACC	60 s 94 °C 60 s 60 °C 2.5 min 72 °C	814 bp
icaB	Forward: 5′-ATGGCTTAAAGCACACGACGC Reverse: 5′-TATCGGCATCTGGTGTGACAG	60 s 94 °C 60 s 59 °C 2.5min 72 °C	526 bp
icaC	Forward: 5′-ATAAACTTGAATTAGTGTATT Reverse: 5′-ATATATAAAACTCTCTTAACA	60 s 94 °C 60 s 45 °C 2.5 min 72 °C	989 bp
icaD	Forward: 5′-AGGCAATATCCAACGGTAA Reverse: 5′-GTCACGACCTTTCTTATATT	60 s 94 °C 60 s 59 °C 2.5 min 72 °C	282 bp

The composition of the reaction mixture for one 50 μL sample included 5 μL amplification buffer, 2.5 μL primer 1, 2.5 μL primer 2, 3.0 μL MgCl2, 1 μL dNTPs, 0.5 μL Taq Polymerase (5 U/μL), 5 μL template DNA, 30.5 μL H2O. The *icaA*, *icaB*, *icaC*, and *icaD* gene amplification products were detected with 2% agarose gel electrophoresis [48]. The positive result was manifested by the presence of a band of appropriate size in each gel for each gene. To determine the position of the amplification product, a DNA size marker was used: ΦX174 DNA/BsuRI (MBI, Fermentas, Lithuania) [51,52].

Data analysis was performed using SPSS 13 (SPSS Inc., Chicago, Illinois, United States of America). The independent samples *t*-test was used to compare differences between the groups. A $p \leq 0.05$ was considered statistically significant.

The study was conducted according to the guidelines of the Declaration of Helsinki and approved by the Bioethical Committee of Medical University of Lublin.

3. Results

3.1. Results of Microbiological Tests

Out of 99 samples, positive cultures were found in 38 samples (38.38%), whereas no bacteria were grown in the remaining 61 samples (61.62%). The average degree of hearing loss in the group of children with positive bacterial culture was 35.9 dB, while in the group with negative bacterial culture it was 25.9 dB ($p = 0.0008$) (Table 4).

Table 4. Degree of hearing impairment depending on the culture result.

Type of Bacteria	Mean Value-M		Standard Deviation-SD
Positive culture	35.9 dB		13.3 dB
Negative culture	25.9 dB		11.5 dB
t		5.659	
p		0.0008	
d		0.801	

t—standard error. p—calculated probability. d—effect size.

PCR reaction confirmed the presence of bacteria in the exudate collected from the middle ear of patients with OME. In the outer reaction, bacterial DNA was confirmed in 19 cases (19.19%), while in the nested reaction, it was found in 42 cases (42.42%).

Opportunistic bacteria were detected much more frequently (76.32%) than pathogenic (23.68%) (Table 5).

Table 5. Culture results.

Type of Bacteria	Percentage
Pathogenic	23.68
Opportunistic	76.32
Total	100.00

Among the pathogenic bacteria *Haemophilus influenzae* was most frequently isolated (38.89%) in our study. Other cultured pathogens were identified as *Streptococcus pneumoniae* (33.33%) and *Staphylococcus aureus MSSA* (27.78%) (Table 6).

Table 6. Type and percentage of pathogenic bacteria in exudate collected from the middle ear space.

Type of Pathogenic Bacteria	Percentage
Haemophilus influenzae	38.89
Streptococcus pneumoniae	33.33
Staphylococcus aureus MSSA	27.78
Total	100.00

Amid the opportunistic bacteria, *Staphylococus* spp. (74.14%) and *Diphtheroids* (20.69%) predominated in the exudate collected from the middle ear space. In addition, *Streptococcus viridans* (3.45%) and *Neisseria* spp. (1.72%) were also found (Table 7).

Table 7. Type and percentage of opportunistic bacteria in exudate collected from the middle ear space.

Type of Opportunistic Bacteria	Percentage
Staphylococcus spp.	74.14
Diphtheroids	20.69
Streptococcus viridans	3.45
Neisseria spp.	1.72
Total	100.00

Based on our study, the type of cultured bacteria has a significant influence on the degree of hearing loss in children ($p = 0.0192$). The average degree of hearing loss in the group of children with pathogenic bacteria found in their exudate was higher (32.3 dB) than in the group with opportunistic bacteria (27.5 dB) (Table 8).

Table 8. Degree of hearing impairment depending on the type of cultured bacteria.

Type of Bacteria	Mean Value-M		Standard Deviation-SD
Pathogenic bacteria	32.3 dB		12.0 dB
Opportunistic bacteria	27.5 dB		12.1 dB
t [1]		3.025	
p [2]		0.0192	
d [3]		0.395	

[1] t—standard error. [2] p—calculated probability. [3] d—effect size.

3.2. Bacterial Biofilm Formation Results

Due to the large amount of *Staphylococcus* spp. bacteria in the culture (74.14%), an attempt was made to analyze the phenotypic and genotypic ability of *Staphylococcus* spp. strains to form biofilm. Out of all *Staphylococcus* spp. isolates, the Api Staph study showed that the most commonly isolated bacteria were *S. epidermidis* (50%). Next predominating organisms were *S. aureus* (18.75%) and *S. sciuri* (12.5%). The remaining strains were identified as: *S. capitis* (6.25%), *S. caprae* (6.25%), and *S. verneri* (6.25%) (Table 9).

Table 9. Type and percentage of *Staphylococcus* spp.

Type of Bacteria	Percentage
Staphylococcus epidermidis	50.00
Staphylococcus aureus	18.75
Staphylococcus sciuri	12.50
Staphylococcus capitis	6.25
Staphylococcus caprae	6.25
Staphylococcus verneri	6.25

Out of all *Staphylococcuss* spp. isolates, 37.5% showed a phenotypic ability to form biofilm, as confirmed by CRA and TCP methods. In addition, 12.5% of *S. epidermidis* strains were *icaA*, *icaB*, *icaC*, and *icaD* positive and 6.25% of *S. epidermidis* strains were only *icaA* and *icaD* positive in PCR. However, 18.75% of *S. epidermidis* were also positive in the TCP and CRA methods, which confirms their biofilm formation potential.

Lastly, 80% of *S. aureus* strains, detected by PCR analysis as *icaA* or *icaB* positive, showed a biofilm negative phenotype in the TCP and CRA methods.

4. Discussion

Research by Bluestone et al. [17] showed that the most common pathogens in ear infections were *Streptoccocus pneumoniae*, *Haemophilus influenzae*, and *Moraxella catarrhalis*. However, the authors found a different percentage of these bacteria in acute otitis media (AOM) and OME. In AOM, the most common bacterium was *Streptococcus pneumoniae*, isolated in 35% of cases, whereas in OME it was found in only 7% of cases. Contrarily, in OME the most frequently isolated bacterium was *Haemophilus influenzae* (15% of exudates) and the second most frequent bacterium was *Moraxella catarrhalis*, cultured in 10% exudates. This study was conducted on the exudate collected from 4589 ears and it is often a model for other researchers. Bluestone did not obtain bacterial growth in 30% of samples, whereas in our study no bacteria were cultured in 61.62% of the samples. Other bacteria considered to be non-pathogenic constituted 45% in Bluestone's research, while in our study opportunistic bacteria were found in 76.32% of positive culture samples. In addition to the three major pathogens, Bluestone et al. identified *Streptococcus aureus* 2%, *Streptococci group A* 1%, *Streptococci alpha* 3%, and *Pseudomonas aeruginosa* [53] 2%. In our study, out of 99 samples collected from middle ear space, *Haemophilus influenzae* was detected in 3.53% of cases, *Streptococcus pneumoniae* in 3% of samples, *Staphyloccocus aureus MSSA* (Methicillin-sensitive *Staphylococcus aureus*) in 2.52% of cases, and *Streptoccocus viridans* in 1% of samples. The most commonly cultured microorganism in our study was *Staphylococcus* spp. (21.71%), and no growth of *Moraxella catarrhalis* was found in the culture. Comparable results were obtained by Park et al. [52], who confirmed the presence of *Haemophilus influenzae* in 7.9% of cases and *Streptococcus pneumonae* in 1.4% of cases, while *Moraxella catarrhalis* was not grown. On the other hand, Poetker et al. [22] reported that the most frequently isolated *Staphylococcus species NOS* (not otherwise specified) was identified in 38 samples from 148 (25.7%). The decrease in the percentage of major pathogens compared with Bluestone's studies, especially reduction of the amount of *Streptococcus pneumoniae*, may be associated with the usage of pneumococcal vaccine. Similar observations are also noted by other authors [54,55].

The positive identification of bacteria by PCR suggests a bacterial etiology for OME. Palmu et al. identified *S. pneumoniae* in 47.1% of middle ear effusion using PCR, compared with 27.3% using standard cultures [56]. In studies of Park et al. [52], bacteria in culture were detected in 14% of cases, while using PCR techniques, bacterial DNA was isolated in 36.7% [52]. Similar results were obtained in research by Choi et al. [57]. In our study, bacteria in culture were detected in 38.38% of cases, whereas by PCR in 61.62% of samples, using outer and nested technique (19.19% and 42.42%, respectively). The authors discuss such a low bacterial detection rate in cultures, explaining this phenomenon with antibiotic therapy before ear drainage, the presence of secretory immunoglobulins, and lysozyme in the middle ear secretion inhibiting bacterial growth, as well as the presence of bacteria in the middle ear in the form of biofilms [52,58].

OME can cause hearing impairment, and in our study the hearing loss varied in the range of 25–40 dB. What is interesting is that the degree of hearing loss was significantly higher in patients from which positive bacterial cultures were obtained. The difference in the audiometric test was on average 10 dB and was statistically significant ($p = 0.0008$). Thus, in the ears with a positive culture, hearing loss was 35.1 dB on average. A positive correlation was also observed between the degree of hearing loss and the presence of pathogenic bacteria as compared with non-pathogenic bacteria. In the case of pathogenic bacteria, the degree of hearing loss was greater ($p = 0.0192$). To the best of our knowledge, no study has been published showing the correlation between the degree of hearing loss and the type of bacteria found in exudate. However, experimental studies could provide some possible reasons why the degree of hearing loss differs between positive and negative cultures. Stenqvist et al. studied electrophysiological changes in the albino rat following instillation of *Pseudomonas aeruginosa* exotoxin A into the middle ear cavity [59]. They found that *Pseudomonas aeruginosa* exotoxin A causes middle ear inflammation, facilitating penetration to the inner ear and that this toxin also reversibly affects cochlear function—*Pseudomonas aeruginosa* exotoxin A raised the ABR threshold over the whole frequency range by 5–25 dB [59], which can explain why in our study the degree of hearing loss was significantly greater in the children in which bacteria were detected.

S. epidermidis is the main organism isolated from foreign material related infections (FMRI), such as infected prosthetic joints, central venous catheters, cerebrospinal fluid shunts, intracardiac devices, artificial heart valves, and vascular grafts [60]. In our study, *S. epidermis* had proven to be the most common microorganism of *Staphylococcus species* to form biofilm in OME. Daniel et al. examined bacterial involvement in OME using confocal laser scanning microscopy (CLSM) and bacterial viability stain [61]. They noticed that among the CLSM-positive samples, 49.0% contained biofilms, and the most common pathogen to form biofilm was *Psudomonas* spp. (4.8%). They also found that coagulase-negative *staphylococci* (CoNS) dominated in the culture (12.9%), and out of all CoNS strains, *S. epidermidis* (3.2%) and *S. lugdunensis* (3.2%) were the most common, similar to our findings. However, among CoNS strains, 25% were able to form biofilm based on their findings [61]. In our study, 37.5% of all *Staphylococcuss* spp. strains showed a phenotypic ability to form biofilm.

It is likely that freely drifting, planktonic bacteria are far less common than those associated with biofilms. The biofilm explains the presence of metabolically active bacteria as well as bacterial endotoxins, despite negative cultures from the middle ear.

The OME etiology model is a chronic middle ear effusion as a result of biofilm formation from pathogenic bacteria on the middle ear mucosa. In this theory, OME is an active chronic bacterial disease rather than an aseptic inflammatory process. Experimental studies seem to confirm this theory because biofilm was found on the mucous membrane of the middle ear of chinchilla in otitis media experimentally induced by *Haemophilus influenzae* [62]. The mucosal biofilm model can explain the observation that metabolically active bacteria are present in negative cultures of OME exudates and that antibiotics are ineffective, while tympanostomy drainage is effective in the treatment of OME [63,64].

The unsatisfactory effect of OME antibiotic therapy may result not only from the genetic resistance of bacteria but also from the slowdown of their metabolism, independent of genetic conditions. The fact that antibiotics can make bacteria difficult to grow explains the postulate that antibiotic stress induces these pathogens to form biofilm [63,64].

Studies on the pharmacokinetics of orally administered antibiotics demonstrate that the killer-mediating drug concentration in vivo is readily available in the middle ear space. However, biofilm-forming bacteria are hundreds of times more resistant to antibiotics than planktonic bacteria, mainly due to the fact that mature parts of the biofilm grow slowly and therefore less frequently interact with the middle ear environment. Thus, the biofilm pattern can explain clinical observations that antibiotics are ineffective in the treatment of OME [63,64].

The biofilm hypothesis is also consistent with clinical observations that ventilation drainage is the most effective method of OME treatment. The non-ventilated middle ear is an ideal environment for biofilm formation because previous viral infections and persistent hypoxia disrupt the normal defense mechanisms of the mucous membrane. The healthy middle ear mucosa consists of ciliated epithelial cells that are involved in the bacterial cell purification mechanisms. It has been proven that the epithelium of the middle ear in OME is deprived of cilia, while rich in secretory cells (their number increases during OME). Placement of the tympanostomy tube restores ventilation of the middle ear and causes an increase in the partial pressure of oxygen, changing the biofilm phenotype. Suction of exudate breaks and reduces the mass of biofilm, increases the oxygen level, and leads to renewal of the ciliary epithelium [65].

However, the following limitations should be noted. Firstly, due to the age of the patients and the lack of cooperation between audiologists and patients younger than 7 years of age, the pure-tone threshold audiometry was possible only in half of the patients (53.95%). In the remaining children, only impedance audiometry and otoacoustic emissions were tested. In our next study we would like to include only patients older than 7 years to be able to determine hearing thresholds in all patients. Moreover, in our study there was no control group of healthy children without any otological problems; however, in future research we will include a control group and compare the microbiological profile between the study and the control group.

5. Conclusions

The obtained results allow the following conclusions to be drawn:

1. *Staphylococcus* spp. in OME may form biofilms, which can explain the chronic character of the disease and negative culture results.

2. Pathogenic bacteria typical of upper respiratory tract infections (*Haemophilus influenzae*, *Streptococcus pneumoniae* and *Staphylococcus aureus MSSA*) as well as opportunistic bacteria (*Staphylococus* spp., *Diphtheroids*, *Streptococcus viridans* and *Neisseria* spp.) may be involved in the etiopathogenesis of otitis media with effusion.

3. The degree of hearing loss correlated with the presence of bacteria, as evidenced by the results of microbiological tests. The degree of hearing loss was significantly higher in patients from which positive bacterial cultures were obtained. The difference in the audiometric test was on average 10 dB and was statistically significant (p = 0.0008).

Better understanding of the pathogens involved in otitis media with effusion development will help to identify high-risk patients and to explain the pathogenesis of the disease. This, in turn, will provide adequate opportunities for the design and implementation of diagnostic tests and effective therapeutic strategies for otitis media with effusion. Hearing can be monitored in patients with positive bacterial cultures and therefore a permanent hearing loss due to otitis media with effusion can be avoided.

Author Contributions: Conceptualization, A.N., L.P.C., and T.S.; methodology, A.N., L.P.C., and T.S.; software, A.N., L.P.C., and T.S.; validation, A.N., L.P.C., and T.S.; formal analysis, A.N., L.P.C., and T.S.; investigation, A.N., L.P.C., and T.S.; resources, A.N., L.P.C., and T.S.; data curation, A.N., L.P.C., and T.S.; writing—original draft preparation, A.N., L.P.C., and T.S.; writing—review and editing, A.N., L.P.C., and T.S.; visualization, A.N., L.P.C., and T.S.; supervision, A.N.; project administration, A.N., L.P.C., and T.S.; funding acquisition—this research received no external funding. All authors have read and agreed to the published version of the manuscript.

Funding: This research received no external funding.

Institutional Review Board Statement: The study was conducted according to the guidelines of the Declaration of Helsinki and approved by the Ethics Committee of Medical University of Lublin (KE-0254/14/2009).

Informed Consent Statement: Informed consent was obtained from all subjects involved in the study.

Data Availability Statement: The data presented in this study are available on request from the corresponding author. The data are not publicly available due to research participants' privacy.

Acknowledgments: Authors have no financial relationships relevant to this article to disclose. This research received no specific grant from any funding agency, commercial, or not-for-profit sectors. No funding was secured for this study. Authors have no competing interests to disclose.

Conflicts of Interest: The authors declare no conflict of interest.

References

1. Ashoor, A. Middle ear effusion in children: Review of recent literature. *J. Fam. Community Med.* **1994**, *1*, 12–18.
2. Bennett, K.E.; Haggard, M.P.; Silva, P.A.; Stewart, I.A. Behaviour and developmental effects of otitis media with effusion into the teens. *Arch. Dis. Child.* **2001**, *85*, 91–95. [CrossRef]
3. Rosenfeld, R.M.; Schwartz, S.R.; Pynnonen, M.A.; Tunkel, D.E.; Hussey, H.M.; Fichera, J.S.; Grimes, A.M.; Hackell, J.M.; Harrison, M.F.; Haskell, H. Clinical Practice Guideline: Tympanostomy Tubes in Children. *Otolaryngol. Head Neck Surg.* **2013**, *149*, S1–S35. [CrossRef]
4. Skoloudik, L.; Kalfert, D.; Valenta, T.; Chrobok, V. Relation between adenoid size and otitis media with effusion. *Eur. Ann. Otorhinolaryngol. Head Neck Dis.* **2018**, *135*, 399–402. [CrossRef]
5. Rosenfeld, R.M.; Shin, J.J.; Schwartz, S.R.; Coggins, R.; Gagnon, L.; Hackell, J.M.; Hoelting, D.; Hunter, L.L.; Kummer, A.W.; Payne, S.C.; et al. Clinical Practice Guideline: Otitis Media with Effusion (Update). *Otolaryngol. Head Neck Surg.* **2016**, *154*, S1–S41. [CrossRef] [PubMed]
6. Elmagd, E.A.A.; Saleem, T.H.; Khalefa, M.E.; Elhawary, B. Pathogenesis and Microbiology of Otitis Media with Effusion in Children. *Int. J. Otolaryngol. Head Neck Surg.* **2019**, *08*, 113–120. [CrossRef]
7. Chonmaitree, T.; Revai, K.; Grady, J.J.; Clos, A.; Patel, J.A.; Nair, S.; Fan, J.; Henrickson, K.J. Viral Upper Respiratory Tract Infection and Otitis Media Complication in Young Children. *Clin. Infect. Dis.* **2008**, *46*, 815–823. [CrossRef] [PubMed]
8. Venekamp, R.P.; Burton, M.J.; van Dongen, T.M.; van der Heijden, G.J.; van Zon, A.; Schilder, A.G. Antibiotics for otitis media with effusion in children. *Cochrane Database Syst. Rev.* **2016**, *2016*, CD009163. [CrossRef] [PubMed]
9. Ahn, J.H.; Kang, W.S.; Kim, J.H.; Koh, K.S.; Yoon, T.H. Clinical manifestation and risk factors of children with cleft palate receiving repeated ventilating tube insertions for treatment of recurrent otitis media with effusion. *Acta Otolaryngol.* **2012**, *132*, 702–707. [CrossRef] [PubMed]
10. Galletti, B.; Gazia, F.; Freni, F.; Nicita, R.A.; Bruno, R.; Galletti, F. Chronic otitis media associated with cholesteatoma in a case of the say-barber-biesecker-young-simpson variant of ohdo syndrome. *Am. J. Case Rep.* **2019**, *20*, 175–178. [CrossRef]
11. Boonacker, C.W.B.; Rovers, M.M.; Browning, G.G.; Hoes, A.W.; Schilder, A.G.M.; Burton, M.J. Adenoidectomy with or without grommets for children with otitis media: An individual patient data meta-analysis. *Health Technol. Assess.* **2014**, *18*, 5. [CrossRef]
12. Miura, M.S.; Mascaro, M.; Rosenfeld, R.M. Association between otitis media and gastroesophageal reflux: A systematic review. *Otolaryngol. Head Neck Surg.* **2012**, *146*, 345–352. [CrossRef]
13. Straetemans, M.; Wiertsema, S.P.; Sanders, E.A.M.; Rijkers, G.T.; Graamans, K.; van der Baan, B.; Zielhuis, G.A. Immunological status in the aetiology of recurrent otitis media with effusion: Serum immunoglobulin levels, functional mannose-binding lectin and Fc receptor polymorphisms for IgG. *J. Clin. Immunol.* **2005**, *25*, 78–86. [CrossRef] [PubMed]
14. Luong, A.; Roland, P.S. The Link Between Allergic Rhinitis and Chronic Otitis Media with Effusion in Atopic Patients. *Otolaryngol. Clin. North. Am.* **2008**, *41*, 311–323. [CrossRef]
15. Kiris, M.; Muderris, T.; Kara, T.; Bercin, S.; Cankaya, H.; Sevil, E. Prevalence and risk factors of otitis media with effusion in school children in Eastern Anatolia. *Int. J. Pediatr. Otorhinolaryngol.* **2012**, *76*, 1030–1035. [CrossRef]
16. Shishegar, M.; Faramarzi, A.; Kazemi, T.; Bayat, A.; Motamedifar, M. Polymerase chain reaction, bacteriologic detection and antibiogram of bacteria isolated from otitis media with effusion in children, Shiraz, Iran. *Iran. J. Med. Sci.* **2011**, *36*, 273–280. [PubMed]
17. Bluestone, C.D.; Stephenson, J.S.; Martin, L.M. Ten-year review of otitis media pathogens. *Pediatr. Infect. Dis. J.* **1992**, *11*, 7–11. [CrossRef]
18. Greiner, O.; Day, P.J.R.; Altwegg, M.; Nadal, D. Quantitative detection of Moraxella catarrhalis in nasopharyngeal secretions by real-time PCR. *J. Clin. Microbiol.* **2003**, *41*, 1386–1390. [CrossRef]
19. Karhdag, T.; Kutbettin, D.; Kaygusuz, I.; Ozden, M.; Yalcin, S.; Ozturk, L. Resistant bacteria in the adenoid tissues of children with Otitis media with effusion. *Int. J. Pediatr. Otorhinolaryngol.* **2002**, *64*, 35–40. [CrossRef]
20. Miller, M.B.; Koltai, P.J.; Hetherington, S.V. Bacterial antigens and neutrophil granule proteins in middle-ear effusions. *Arch. Otolaryngol. Neck Surg.* **1990**, *116*, 335–337. [CrossRef] [PubMed]
21. Murphy, T.F.; Kirkham, C. Biofilm formation by nontypeable Haemophilus influenzae: Strain variability, outer membrane antigen expression and role of pili. *BMC Microbiol.* **2002**, *2*, 1–8. [CrossRef] [PubMed]
22. Poetker, D.M.; Lindstrom, D.R.; Edmiston, C.E.; Krepel, C.J.; Link, T.R.; Kerschner, J.E. Microbiology of middle ear effusions from 292 patients undergoing tympanostomy tube placement for middle ear disease. *Int. J. Pediatr. Otorhinolaryngol.* **2005**, *69*, 799–804. [CrossRef] [PubMed]

23. Smith-Vaughan, H.; Byun, R.; Nadkarni, M.; Jacques, N.A.; Hunter, N.; Halpin, S.; Morris, P.S.; Leach, A.J. Measuring nasal bacterial load and its association with otitis media. *BMC Ear Nose Throat Disord.* **2006**, *6*, 10. [CrossRef] [PubMed]
24. Stenfors, L.E.; Raisanen, S. Occurrence of Streptococcus pneumoniae and Haemophilus influenzae in otitis media with effusion. *Clin. Otolaryngol. Allied Sci.* **1992**, *17*, 195–199. [CrossRef]
25. Ulualp, S.O.; Sahin, D.; Yilmaz, N.; Anadol, V.; Peker, O.; Gursan, O. Increased adenoid mast cells in patients with otitis media with effusion. *Int. J. Pediatr. Otorhinolaryngol.* **1999**, *49*, 107–114. [CrossRef]
26. Aly, B.H.; Hamad, M.S.; Mohey, M.; Amen, S. Polymerase Chain Reaction (PCR) Versus Bacterial Culture in Detection of Organisms in Otitis Media with Effusion (OME) in Children. *Indian J. Otolaryngol. Head Neck Surg.* **2012**, *64*, 51–55. [CrossRef]
27. Beswick, A.J.; Lawley, B.; Fraise, A.P.; Pahor, A.L.; Brown, N.L. Detection of Alloiococcus otitis in mixed bacterial populations from middle-ear effusions of patients with otitis media. *Lancet* **1999**, *354*, 386–389. [CrossRef]
28. Hendolin, P.H.; Kärkkäinen, U.; Himi, T.; Markkanen, A.; Ylikoski, J. High incidence of Alloiococcus otitis in otitis media with effusion. *Pediatr. Infect. Dis. J.* **1999**, *18*, 860–865. [CrossRef]
29. Belfield, K.; Bayston, R.; Birchall, J.P.; Daniel, M. Do orally administered antibiotics reach concentrations in the middle ear sufficient to eradicate planktonic and biofilm bacteria? A review. *Int. J. Pediatr. Otorhinolaryngol.* **2015**, *79*, 296–300. [CrossRef] [PubMed]
30. Davies, D.G.; Parsek, M.R.; Pearson, J.P.; Iglewski, B.H.; Costerton, J.W.; Greenberg, E.P. The involvement of cell-to-cell signals in the development of a bacterial biofilm. *Science* **1998**, *280*, 295–298. [CrossRef] [PubMed]
31. Dingman, J.R.; Rayner, M.G.; Mishra, S.; Zhang, Y.; Ehrlich, M.D.; Post, J.C.; Ehrlich, G.D. Correlation between presence of viable bacteria and presence of endotoxin in middle-ear effusions. *J. Clin. Microbiol.* **1998**, *36*, 3417–3419. [CrossRef] [PubMed]
32. Ehrlich, G.D.; Veeh, R.; Wang, X.; William Costerton, J.; Hayes, J.D.; Hu, F.Z.; Daigle, B.J.; Ehrlich, M.D.; Christopher Post, J. Mucosal biofilm formation on middle-ear mucosa in the chinchilla model of otitis media. *J. Am. Med. Assoc.* **2002**, *287*, 1710–1715. [CrossRef]
33. Erlandsen, S.L.; Kristich, C.J.; Dunny, G.M.; Wells, C.L. High-resolution visualization of the microbial glycocalyx with low-voltage scanning electron microscopy: Dependence on cationic dyes. *J. Histochem. Cytochem.* **2004**, *52*, 1427–1435. [CrossRef] [PubMed]
34. Polachek, A.; Greenberg, D.; Lavi-Givon, N.; Broides, A.; Leiberman, A.; Dagan, R.; Leibovitz, E. Relationship among peripheral leukocyte counts, etiologic agents and clinical manifestations in acute otitis media. *Pediatr. Infect. Dis. J.* **2004**, *23*, 406–413. [CrossRef] [PubMed]
35. Rayner, M.G.; Zhang, Y.; Gorry, M.C.; Chen, Y.; Post, J.C.; Ehrlich, G.D. Evidence of bacterial metabolic activity in culture-negative otitis media with effusion. *J. Am. Med. Assoc.* **1998**, *279*, 296–299. [CrossRef]
36. Galli, J.; Ardito, F.; Calò, L.; Mancinelli, L.; Imperiali, M.; Parrilla, C.; Picciotti, P.M.; Fadda, G. Recurrent upper airway infections and bacterial biofilms. *J. Laryngol. Otol.* **2007**, *121*, 341–344. [CrossRef]
37. Zuliani, G.; Carron, M.; Gurrola, J.; Coleman, C.; Haupert, M.; Berk, R.; Coticchia, J. Identification of adenoid biofilms in chronic rhinosinusitis. *Int. J. Pediatr. Otorhinolaryngol.* **2006**, *70*, 1613–1617. [CrossRef]
38. Śliwa, L.; Hatzopoulos, S.; Kochanek, K.; Piłka, A.; Senderski, A.; Skarzyński, P.H. A comparison of audiometric and objective methods in hearing screening of school children. A preliminary study. *Int. J. Pediatr. Otorhinolaryngol.* **2011**, *75*, 483–488. [CrossRef] [PubMed]
39. Kemp, D.T. Otoacoustic emissions, their origin in cochlear function, and use. *Br. Med. Bull.* **2002**, *63*, 223–241. [CrossRef]
40. Musiek, F.E.; Shinn, J.; Chermak, G.D.; Bamiou, D.E. Perspectives on the pure-tone audiogram. *J. Am. Acad. Audiol.* **2017**, *28*, 655–671. [CrossRef]
41. NHS. Bacteriology: Catalase Test. *UK Stand. Microbiol. Investig.* **2019**, *8*, 1–27.
42. Smole, S.C.; Aronson, E.; Durbin, A.; Brecher, S.M.; Arbeit, R.D. Sensitivity and Specificity of an Improved Rapid Latex Agglutination Test for Identification of Methicillin-Sensitive and -Resistant Staphylococcus aureus Isolates. *J. Clin. Microbiol.* **1998**, *36*, 1109–1112. [CrossRef]
43. Chamberlain, N. Coagulase Test for Staphylococcus Species. *Am. Soc. Microbiol.* **2009**, 1–4.
44. Chapin, K.; Musgnug, M. Evaluation of three rapid methods for the direct identification of Staphylococcus aureus from positive blood cultures. *J. Clin. Microbiol.* **2003**, *41*, 4324–4327. [CrossRef]
45. Pinto, T.C.A.; Souza, A.R.V.; De Pina, S.E.C.M.; Costa, N.S.; Neto, A.A.B.; Neves, F.P.G.; Merquior, V.L.C.; Dias, C.A.G.; Peralta, J.M.; Teixeira, L.M. Phenotypic and molecular characterization of optochin-resistant streptococcus pneumoniae isolates from Brazil, with description of five novel mutations in the atpC gene. *J. Clin. Microbiol.* **2013**, *51*, 3242–3249. [CrossRef]
46. Arbique, J.C.; Poyart, C.; Trieu-Cuot, P.; Quesne, G.; Carvalho, M.D.G.S.; Steigerwalt, A.G.; Morey, R.E.; Jackson, D.; Davidson, R.J.; Facklam, R.R. Accuracy of phenotypic and genotypic testing for identification of Streptococcus pneumoniae and description of Streptococcus pseudopneumoniae sp. nov. *J. Clin. Microbiol.* **2004**, *42*, 4686–4696. [CrossRef]
47. Gok, U.; Bulut, Y.; Keles, E.; Yalcin, S.; Doymaz, M.Z. Bacteriological and PCR analysis of clinical material aspirated from otitis media with effusions. *Int. J. Pediatr. Otorhinolaryngol.* **2001**, *60*, 49–54. [CrossRef]
48. Lee, P.Y.; Costumbrado, J.; Hsu, C.Y.; Kim, Y.H. Agarose gel electrophoresis for the separation of DNA fragments. *J. Vis. Exp.* **2012**, *62*, 3923. [CrossRef] [PubMed]
49. Hassan, A.; Usman, J.; Kaleem, F.; Omair, M.; Khalid, A.; Iqbal, M. Evaluation of different detection methods of biofilm formation in the clinical isolates. *Brazilian J. Infect. Dis.* **2011**, *15*, 305–311. [CrossRef]

50. Ziebuhr, W.; Krimmer, V.; Rachid, S.; Lößner, I.; Götz, F.; Hacker, J. A novel mechanism of phase variation of virulence in Staphylococcus epidermidis: Evidence for control of the polysaccharide intercellular adhesin synthesis by alternating insertion and excision of the insertion sequence element IS256. *Mol. Microbiol.* **1999**, *32*, 345–356. [CrossRef] [PubMed]
51. De Silva, G.D.I.; Kantzanou, M.; Justice, A.; Massey, R.C.; Wilkinson, A.R.; Day, N.P.J.; Peacock, S.J. The ica operon and biofilm production in coagulase-negative staphylococci associated with carriage and disease in a neonatal intensive care unit. *J. Clin. Microbiol.* **2002**, *40*, 382–388. [CrossRef]
52. Park, C.-W.; Han, J.-H.; Jeong, J.-H.; Cho, S.-H.; Kang, M.-J.; Tae, K.; Lee, S.-H. Detection rates of bacteria in chronic otitis media with effusion in children. *J. Korean Med. Sci.* **2004**, *19*, 735–738. [CrossRef]
53. Galletti, F.; Cammaroto, G.; Galletti, B.; Quartuccio, N.; Di Mauro, F.; Baldari, S. Technetium-99m (99mTc)-labelled sulesomab in the management of malignant external otitis: Is there any role? *Eur. Arch. Oto Rhino Laryngol.* **2015**, *272*, 1377–1382. [CrossRef]
54. Black, S.; Shinefield, H.; Fireman, B.; Lewis, E.; Ray, P.; Hansen, J.R.; Elvin, L.; Ensor, K.M.; Hackell, J.; Siber, G.; et al. Efficacy, safety and immunogenicity of heptavalent pneumococcal conjugate vaccine in children. Northern California Kaiser Permanente Vaccine Study Center Group. *Pediatr. Infect. Dis. J.* **2000**, *19*, 187–195. [CrossRef]
55. Caspary, H.; Welch, J.C.; Lawson, L.; Darrow, D.; Buescher, S.; Shahab, S.; Derkay, C.S. Impact of pneumococcal polysaccharide vaccine (Prevnar) on middle ear fluid in children undergoing tympanostomy tube insertion. *Laryngoscope* **2004**, *114*, 975–980. [CrossRef]
56. Palmu, A.I.; Saukkoriipi, P.A.; Lahdenkari, M.I.; Kuisma, L.K.; Makela, P.H.; Kilpi, T.M.; Leinonen, M. Does the presence of pneumococcal DNA in middle-ear fluid indicate pneumococcal etiology in acute otitis media? *J. Infect. Dis.* **2004**, *189*, 775–784. [CrossRef]
57. Choi, Y.C.; Park, Y.S.; Yeo, S.W.; Chae, S.; Jung, D.G.; Kim, S.W. Detection of Haemophilus influenzae and Streptococcus pneumoniae by polymerase chain reaction (PCR) in chronic Otitis media with effusion (OME). *Korean J. Otolaryngol.* **1998**, *41*, 846–850.
58. Hall-Stoodley, L.; Hu, F.Z.; Gieseke, A.; Nistico, L.; Nguyen, D.; Hayes, J.; Forbes, M.; Greenberg, D.P.; Dice, B.; Burrows, A.; et al. Direct detection of bacterial biofilms on the middle-ear mucosa of children with chronic otitis media. *J. Am. Med. Assoc.* **2006**, *296*, 202–211. [CrossRef] [PubMed]
59. Stenqvist, M.; Anniko, M.; Pettersson, Å. Effect of Pseudomonas aeruginosa exotocin A on inner ear function. *Acta Otolaryngol.* **1997**, *117*, 73–79. [CrossRef] [PubMed]
60. Büttner, H.; Mack, D.; Rohde, H. Structural basis of Staphylococcus epidermidis biofilm formation: Mechanisms and molecular interactions. *Front. Cell. Infect. Microbiol.* **2015**, *5*, 14. [CrossRef] [PubMed]
61. Daniel, M.; Imtiaz-Umer, S.; Fergie, N.; Birchall, J.P.; Bayston, R. Bacterial involvement in Otitis media with effusion. *Int. J. Pediatr. Otorhinolaryngol.* **2012**, *76*, 1416–1422. [CrossRef]
62. Suzuki, K.; Bakaletz, L.O. Synergistic effect of adenovirus type 1 and nontypeable Haemophilus influenzae in a chinchilla model of experimental otitis media. *Infect. Immun.* **1994**, *62*, 1710–1718. [CrossRef] [PubMed]
63. Fergie, N.; Bayston, R.; Pearson, J.P.; Birchall, J.P. Is otitis media with effusion a biofilm infection? *Clin. Otolaryngol. Allied Sci.* **2004**, *29*, 38–46. [CrossRef] [PubMed]
64. Qureishi, A.; Lee, Y.; Belfield, K.; Birchall, J.P.; Daniel, M. Update on otitis media—Prevention and treatment. *Infect. Drug Resist.* **2014**, *7*, 15–24. [PubMed]
65. Rosenfeld, R. Nonsurgical management of surgical otitis media with effusion. *J. Laryngol. Otol.* **1995**, *109*, 811–816. [CrossRef] [PubMed]

International Journal of
Environmental Research and Public Health

MDPI

Case Report

Multimodal Imaging in Susac Syndrome: A Case Report and Literature Review

Simone Alex Bagaglia [1], Franco Passani [2], Giovanni William Oliverio [3,*], Leandro Inferrera [3], Feliciana Menna [4], Alessandro Meduri [3] and Cosimo Mazzotta [5]

1 Department of Ophthalmology, University of Siena, Policlinico Santa Maria alle Scotte, Viale Mario Bracci, 53100 Siena, Italy; dott.bagaglia@gmail.com
2 Department of Surgery Section of Ophthalmology USL Toscana Nord Ovest, Civic Hospital of Carrara, 54033 Carrara, Italy; f.passani@usl1.toscana.it
3 Biomedical, Dental and Morphological and Functional Images Sciences, Department of Ophthalmology, University of Messina, 98124 Messina, Italy; inferreraleandro@gmail.com (L.I.); ameduri@unime.it (A.M.)
4 Head and Neck Department, University of Naples "Federico II", 80131 Naples, Italy; feliciana.menna@gmail.com
5 Siena Crosslinking Center, 53100 Siena, Italy; cgmazzotta@libero.it
* Correspondence: gioliverio@unime.it

Abstract: Susac syndrome (SS) is a rare microangiopathy that involves arterioles of the brain, retina, and cochlea. Diagnosis is extremely difficult because of the rarity of the disease and because the signs and symptoms often occur at different times. Multidisciplinary approaches and multimodal images are mandatory for diagnosis and prompt therapy. In this report, we describe a case of SS and the application of multimodal retinal imaging to evaluate the ophthalmologic changes and to confirm diagnosis. Early diagnosis and therapy based on the associations of steroids and immunosuppressants are necessary to limit the sequelae of the disease.

Keywords: susac syndrome; multimodal imaging; optical coherence tomography angiography; retinal branch artery occlusion; fluorescein angiography

check for updates

Citation: Bagaglia, S.A.; Passani, F.; Oliverio, G.W.; Inferrera, L.; Menna, F.; Meduri, A.; Mazzotta, C. Multimodal Imaging in Susac Syndrome: A Case Report and Literature Review. *Int. J. Environ. Res. Public Health* **2021**, *18*, 3435. https://doi.org/10.3390/ijerph18073435

Academic Editor: U Rajendra Acharya

Received: 10 March 2021
Accepted: 23 March 2021
Published: 26 March 2021

1. Introduction

Susac syndrome (SS) is a rare autoimmune syndrome characterized by microvascular alterations involving the precapillary arterioles of the brain, retina, cochlea, and semicircular canals [1]. This disease affects mainly women, and the age of onset ranges from 9 to 58 years [2,3]. Laboratory investigations, brain magnetic resonance imaging (MRI), fluorescein angiography (FA), and audiometry findings enable its diagnosis [4]. Diagnosis is extremely difficult as the disease is particularly rare and because the signs and symptoms often occur at different times. Multidisciplinary approaches and multimodal imaging are frequently used to confirm SS diagnosis. Brain magnetic resonance imaging (BMRI) findings are not specific and could be suggestive of several neurological diseases. Moreover, prompt treatment including corticosteroids and immunosuppressants is required to limit complications of the disease [5]. Retinal branch artery occlusion (BRAO), vascular leakage, and arteriolar wall hyper-fluorescence (glass plaques) are the primary ophthalmic findings and are conventionally seen in FA and wide-field color fundus photography. Optical coherence tomography (OCT) improved the evaluation of the retinal layers and may also help differentiate SS from other retinal diseases [6–8]. Optical coherence tomography angiography (OCTA) is a novel noninvasive imaging technique that provides retinal and choroidal volumetric bold flow data, permitting retinal vessel reconstruction for imaging of retinal perfusion and function [9–12].

In this paper, we considered the application of the multimodal retinal imaging approach to describe a case of SS.

2. Case Report

A 33-year-old man was admitted to the emergency unit for headaches, a referred visual field loss, dizziness, and weakness of the limbs. The patient had a history of malignant external otitis [13], hearing loss, and cochlear implant surgery [14–16] and used a mandibular advanced device for obstructive sleep apnea syndrome [17]. Neurological evaluation demonstrated a decrease in attention and memory, ataxia, dysmetria, and mild weakness of the extremities. Although the brain CT scans were normal, MRI revealed numerous rounds and focal lesions in the corpus callosum, subcortical white matter, deep grey matter (basal ganglia and thalami), and cerebellum (Figure 1). Additionally, laboratory analyses were normal while anti-smooth muscle antibodies (ASMA), anti-nuclear antibodies (ANA), and anti-cardiolipin IgM antibodies were positive. Two years ago, the patient underwent photorefractive refractive keratectomy (PRK) for myopia [18]. His best-corrected visual acuity (BCVA) was 20/20 Snellen in both eyes, with spherical equivalent refractions of −0.75 in the right eye and of −0.50 in the left eye. The intraocular pressure was 16 mmHg in both eyes (Goldmann Applanation Tonometer). Anterior segment examination was normal in both eyes; however, fundus evaluation showed ischemic retinal whitening in the inferior-temporal and nasal periphery outside the vascular arcades, related to BRAO. FA was performed by a Heidelberg retina angiograph (HRA II, Heidelberg Engineering, Dossenheim, Germany), showing in both eyes arterial occlusion with vasculitis signs (Figure 2). Finally, considering the clinical, laboratory, serological, and neuro-radiological findings, SS was diagnosed. Further ophthalmologic investigations were conducted; in particular, the patient underwent multimodal retinal imaging including wide-field color fundus photography acquired using a non-mydriatic fundus camera (Canon CR-2, Tokyo, Japan) and high-definition optical coherence tomography angiography (OCTA, Angio OCT Scans, Heidelberg Engineering, Heidelberg, Germany). Color fundus photography confirmed a slight area of retinal ischemia observed in a fundus examination conducted first and revealed the existence of glass plaques as yellowish lipid sediments at the mid-segment of the retinal arterioles (Figure 3). The autofluorescence images did not show specific alterations in physiological retinal autofluorescence. The OCTA scans revealed decreased vascular perfusion in correspondence with the ischemic area, previously observed in FA, as well as an increased foveal avascular zone (FAZ) area in both superficial and deep vascular plexuses (Figure 4). OCT also showed a reduction in the formation of the inner limiting membrane to internal nuclear layer areas.

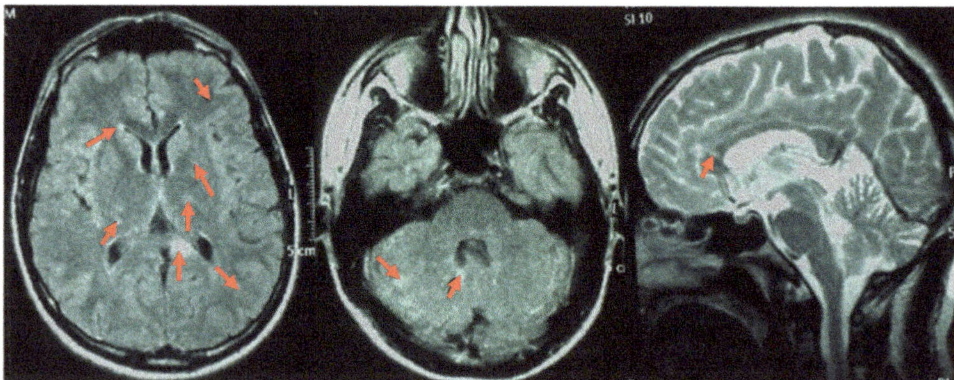

Figure 1. Magnetic resonance of the brain: the red arrows show various rotund and ellipsoid focal lesions involving the corpus callosum, periventricular and subcortical white matter, deep grey matter (basal ganglia and thalami), and cerebellum.

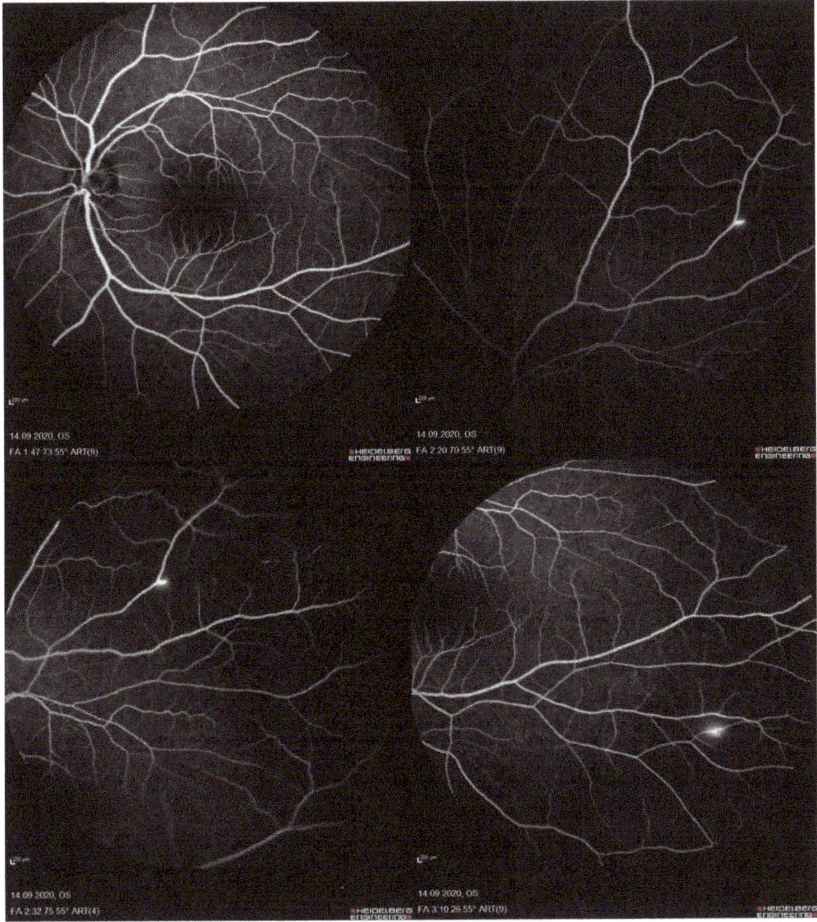

Figure 2. Fluorescein angiography examination exhibiting branch retinal artery occlusion in the left eye.

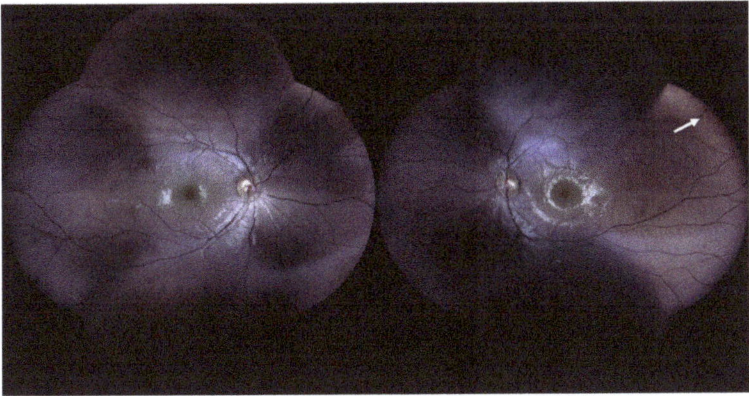

Figure 3. Color fundus photography revealing a glass plaque (white arrow) as yellowish lipid sediments at the mid-segment of the retinal arterioles in the left eye.

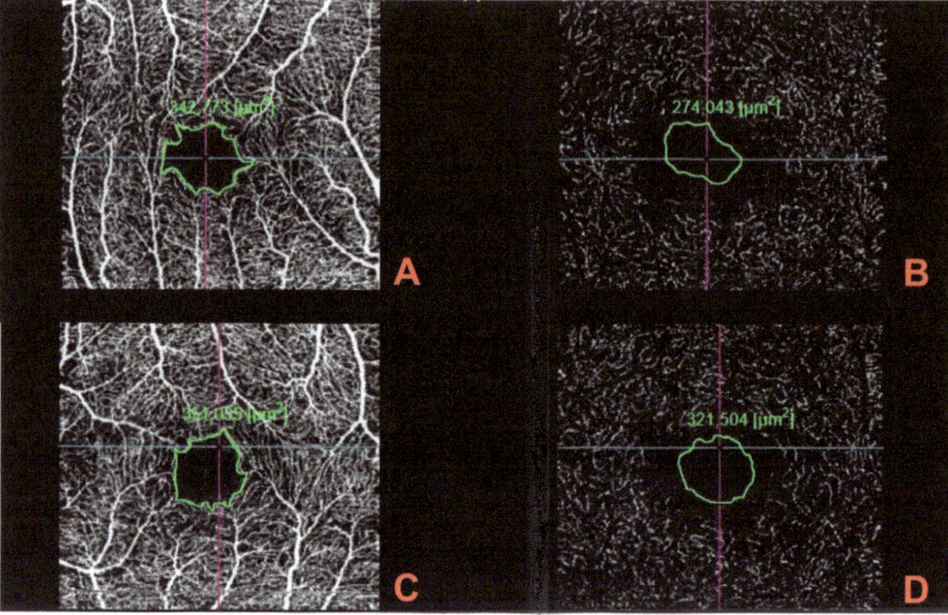

Figure 4. Foveal avascular zone enlargement in optical coherence tomography angiography images of the superficial and deep layers (**A**,**B**) right eye; (**C**,**D**) left eye).

3. Discussion

Although the clinical trio including focal neurological deficits, hearing defeat, and retinal arterial occlusions is considered pathognomonic, the whole clinical manifestation occurs in only 13% of patients at onset [1–3]. The most suggestive neurological symptoms and clinical signs suggestive of brain involvement are changes in consciousness or new cognitive deficiency or behavioral modifications, new focal neurological symptoms, and headaches [18,19]. Susac et al. defined a neuroimaging triad of white matter lesions in the corpus callosum, deep grey matter alterations, and leptomeningeal enhancement. Distinctive neuroimaging alterations, defined as T2/fluid-attenuated inversion recovery hyper-intense multifocal, round brain lesions with at least one centrally set in the corpus callosum, are definitive [18,19]. Additionally, their dimensions and forms are inconstant, including square, triangular, or rectilinear lesions. Certain brain involvement is described as the occurrence of at least one of the clinical manifestations in addition to the characteristic MRI findings [18,19]. These lesions are secondary to arteriolar infarction in the callosum and, afterwards, cavitate and change into manifestation of a hole. Symptoms correlated with retinal ischemia could manifest as a visual field altitudinal defect or central-paracentral scotoma [18]. However, in the case of a far periphery of retina involvement, patients could be asymptomatic. In some patients, the encephalopathy could be severe and visual symptoms may not be referred even though ophthalmologic examination is mandatory if Susac syndrome is suspected, even in asymptomatic patients [19]. The fundus examination may reveal the presence of Gass plaques as yellow reflective lesions due to an autoimmune local reaction in the retinal arterial wall and characteristic retinal whitening in the area of BRAO [20–23]. However, Gass plaques are characteristic but not pathognomonic of Susac syndrome and could be observed in various retinal disorders such as Eale's disease and retinal lymphoma. Furthermore, vascular alteration is common in the early phases of the disease and could then disappear [19]. FA is able to document the vascular wall damage as arteriolar hyper-fluorescence at the site of infarction [4]. MRI and FA remain the bases of diagnostic assessment [18,19]. Spectral-domain optical coherence tomography (SD-OCT) is able to evaluate the microstructure alterations of each individual retinal layer, documenting the ischemic swelling and better analyzing the retinal changes compared to FA, but it does not investigate vascular perfusion [6]. OCTA is a safe, recent, dye-less, and fast instrumental examination able to acquire high-resolution images of retinal microcirculation [10]. In this case, we report the microvascular findings in a patient affected by acute BRAO and correlating the OCTA features with those of wide-field color fundus photography and FA. We highlighted alterations in both superficial and deep retinal vascular plexuses, such as a low flow area, in the course of BRAO and reported a reduction in macula thickness in OCT images in the formation of the inner limiting membrane to internal nuclear layer areas. Some authors reported specific alterations such as a reduction in retinal autofluorescence in the ischemic area [11]; in our case, we did not observe any specific alteration in retinal autofluorescence. Furthermore, SS is an important differential diagnosis in numerous cerebrovascular conditions, and early identification supports starting prompt treatment, reducing relapses, and improving outcomes [5]. Although FA is necessary for identification of the disease, this is a time-consuming, dye-dependent technique and does not allow users to examinate the deep capillary plexus. OCTA may be an effective alternative to the standard FA as it consents to the microvascular analysis of both superficial and deep capillary plexi and to monitoring of the vascular density changes without dye injection [24]. However, additional prospective studies are necessary to establish the role of OCTA, wide-field color fundus photography, and AF to monitor disease activity and to determine the efficacy of new therapeutic approaches [25].

4. Conclusions

In conclusion, MRI and FA are the main diagnostic instruments and may lead clinical ophthalmologists to the correct diagnosis of SS; however, new diagnostic instruments such as OCTA and wide-field color retinal photography may provide an advantage in early diagnosis and follow-up of SS, offering an effective alternative to standard FA; in evaluating microvascular changes in all capillary plexuses; and in monitoring retinal vessel density alterations during prompt therapy without dye injection.

Author Contributions: Conceptualization, F.M., S.A.B. and A.M.; methodology, G.W.O.; validation, F.M., C.M. and F.P.; formal analysis, G.W.O.; investigation, L.I.; resources, S.A.B.; data curation, G.W.O.; writing—original draft preparation, S.A.B.; writing—review and editing, F.M., A.M. All authors have read and agreed to the published version of the manuscript.

Funding: This research received no external funding.

Institutional Review Board Statement: Not applicable.

Informed Consent Statement: Written consent was obtained from patient.

Data Availability Statement: The data used to support the findings of this study are available from the corresponding author upon request.

Conflicts of Interest: No potential conflicts of interest were reported by any author.

References

1. Susac, J.O.; Hardman, J.M.; Selhorst, J.B. Microangiopathy of the brain and retina. *Neurology* **1979**, *29*, 313–316. [CrossRef]
2. Do, T.H.; Fisch, C.; Evoy, F. Susac Syndrome: Report of four cases and review of the literature. *Am. J. Neuroradiol.* **2004**, *25*, 382–383. [PubMed]
3. Eluvathingal Muttikkal, T.J.; Vattoth, S.; Keluth Chavan, V.N. Susac syndrome in a young child. *Pediatr. Radiol.* **2007**, *37*, 710–713. [CrossRef] [PubMed]
4. Martinet, N.; Fardeau, C.; Adam, R.; Bodaghi, B.; Papo, T.; Piette, J.C.; Lehoang, P. Fluorescein and indocyanine green angiographies in Susac syndrome. *Retina* **2007**, *27*, 1238–1242. [CrossRef] [PubMed]
5. Catarsi, E.; Pelliccia, V.; Pizzanelli, C.; Pesaresi, I.; Cosottini, M.; Migliorini, P.; Tavoni, A. Cyclophosphamide and methotrexate in Susac's Syndrome: A successful sequential therapy in a case with involvement of the cerebellum. *Clin. Rheumatol.* **2015**, *34*, 1149–1152. [CrossRef]

6. Agarwal, A.; Soliman, M.; Sarwar, S.; Sadiq, M.A.; Do, D.V.; Nguyen, Q.D.; Sepah, Y.J. Spectral-domain optical coherence tomography evaluation of retinal structure in patients with susacs syndrome. *Retin. Cases Brief Rep.* **2017**, *11*, 123–125. [CrossRef] [PubMed]
7. Ceravolo, I.; Oliverio, G.W.; Alibrandi, A.; Bhatti, A.; Trombetta, L.; Rejdak, R.; Toro, M.D.; Trombetta, C.J. The Application of Structural Retinal Biomarkers to Evaluate the Effect of Intravitreal Ranibizumab and Dexamethasone Intravitreal Implant on Treatment of Diabetic Macular Edema. *Diagnostics (Basel)* **2020**, *10*, 413. [CrossRef]
8. Grenga, P.L.; Fragiotta, S.; Cutini, A.; Meduri, A.; Vingolo, E.M. Enhanced depth imaging optical coherence tomography in adult-onset foveomacular vitelliform dystrophy. *Eur. J. Ophthalmol.* **2016**, *26*, 145–151. [CrossRef] [PubMed]
9. Spaide, R.F.; Klancnik, J.M., Jr.; Cooney, M.J. Retinal vascular layers imaged by fluorescein angiography and optical coherence tomography angiography. *JAMA Ophthalmol.* **2015**, *133*, 45–50. [CrossRef]
10. Mastropasqua, R.; Toto, L.; Senatore, A.; D'Uffizi, A.; Neri, P.; Mariotti, C.; Maccarone, M.T.; Di Antonio, L. Optical coherence tomography angiography findings in Susac's syndrome: A case report. *Int. Ophthalmol.* **2018**, *38*, 1803–1808. [CrossRef]
11. Salvanos, P.; Moe, M.C.; Utheim, T.P.; Bragadóttir, R.; Kerty, E. Ultra-wide-field fundus imaging in the diagnosis and follow-up of susac syndrome. *Retin. Cases Brief Rep.* **2018**, *12*, 234–239. [CrossRef] [PubMed]
12. Giuffrè, C.; Miserocchi, E.; Marchese, A.; Cicinelli, M.V.; Bruschi, E.; Querques, G.; Bandello, F.M.; Modorati, G.M. Widefield OCT angiography and ultra-widefield multimodal imaging of Susac syndrome. *Eur. J. Ophthalmol.* **2019**, *10*, 1120672119843281. [CrossRef] [PubMed]
13. Galletti, F.; Cammaroto, G.; Galletti, B.; Quartuccio, N.; Di Mauro, F.; Baldari, S. Technetium-99m (^{99}mTc)-labelled sulesomab in the management of malignant external otitis: Is there any role? *Eur. Arch. Otorhinolaryngol.* **2015**, *272*, 1377–1382. [CrossRef]
14. Galletti, F.; Freni, F.; Gazia, F.; Galletti, B. Endomeatal approach in cochlear implant surgery in a patient with small mastoid cavity and procident lateral sinus. *BMJ Case Rep.* **2019**, *6*, 12. [CrossRef] [PubMed]
15. Freni, F.; Gazia, F.; Slavutsky, V.; Scherdel, E.P.; Nicenboim, L.; Posada, R.; Portelli, D.; Galletti, B.; Galletti, F. Cochlear Implant Surgery: Endomeatal Approach versus Posterior Tympanotomy. *Int. J. Environ. Res. Public Health* **2020**, *12*, 4187. [CrossRef] [PubMed]
16. Ciodaro, F.; Freni, F.; Mannella, V.K.; Gazia, F.; Maceri, A.; Bruno, R.; Galletti, B.; Galletti, F. Use of 3D Volume Rendering Based on High-Resolution Computed Tomography Temporal Bone in Patients with Cochlear Implants. *Am. J. Case Rep.* **2019**, *12*, 184–188. [CrossRef]
17. Flores-Orozco, E.I.; Tiznado-Orozco, G.E.; Díaz-Peña, R.; Orozco, E.I.F.; Galletti, C.; Gazia, F.; Galletti, F. Effect of a Mandibular Advancement Device on the Upper Airway in a Patient With Obstructive Sleep Apnea. *J. Craniofac. Surg.* **2019**, *31*, e31–e35. [CrossRef]
18. Kleffner, I.; Dörr, J.; Ringelstein, M.; Gross, C.C.; Böckenfeld, Y.; Schwindt, W.; Sundermann, B.; Lohmann, H.; Wersching, H.; Promesberger, J.; et al. Diagnostic criteria for Susac syndrome. *J. Neurol. Neurosurg. Psychiatry* **2016**, *87*, 1287–1295. [CrossRef]
19. Vishnevskia-Dai, V.; Chapman, J.; Sheinfeld, R.; Sharon, T.; Huna-Baron, R.; Manor, R.S.; Shoenfeld, Y.; Zloto, O. Susac syndrome: Clinical characteristics, clinical classification, and long-term prognosis. *Medicine (Baltimore)* **2016**, *95*, e5223. [CrossRef]
20. Brandt, A.U.; Oberwahrenbrock, T.; Costello, F.; Fielden, M.; Gertz, K.; Kleffner, I.; Paul, F.; Bergholz, R.; Dörr, J. Retinal lesion evolution in susac syndrome. *Retina* **2016**, *36*, 366–374. [CrossRef]
21. Zigiotti, G.L.; Cavarretta, S.; Morara, M.; Nam, S.M.; Ranno, S.; Pichi, F.; Lembo, A.; Lupo, S.; Nucci, P.; Meduri, A. Standard enucleation with aluminium oxide implant (bioceramic) covered with patient's sclera. *Sci. World J.* **2012**, *2012*, 481584. [CrossRef] [PubMed]
22. Agarwal, A.; Brubaker, J.W.; Mamalis, N.; Kumar, D.A.; Jacob, S.; Chinnamuthu, S.; Nair, V.; Prakash, G.; Meduri, A.; Agarwal, A. Femtosecond-assisted lamellar keratoplasty in atypical Avellino corneal dystrophy of Indian origin. *Eye Contact Lens* **2009**, *35*, 272–274. [CrossRef] [PubMed]
23. Sauma, J.; Rivera, D.; Wu, A.; Donate-Lopez, J.; Gallego-Pinazo, R.; Chilov, M.; Wu, M.; Wu, L. Susac's syndrome: An update. *Br. J. Ophthalmol.* **2020**, *104*, 1190–1195. [CrossRef] [PubMed]
24. Oliverio, G.W.; Ceravolo, I.; Bhatti, A.; Trombetta, C.J. Foveal avascular zone analysis by optical coherence tomography angiography in patients with type 1 and 2 diabetes and without clinical signs of diabetic retinopathy. *Int. Ophthalmol.* **2021**, *41*, 649–658. [CrossRef]
25. Heng, L.Z.; Bailey, C.; Lee, R.; Dick, A.; Ross, A. A review and update on the ophthalmic implications of Susac syndrome. *Surv. Ophthalmol.* **2019**, *64*, 477–485. [CrossRef]

International Journal of
Environmental Research and Public Health

MDPI

Article

Eustachian Tube Function Assessment after Radiofrequency Turbinate Reduction in Atopic and Non-Atopic Patients

Francesco Martines [1,2,*], Francesco Dispenza [2,3], Federico Sireci [3], Salvatore Gallina [3] and Pietro Salvago [1]

1 Bi.N.D. Department, University of Palermo, Via del Vespro, 129, 90127 Palermo, Italy; pietrosalvago@libero.it
2 Istituto Euromediterraneo di Scienza e Tecnologia—IEMEST, Via M. Miraglia 20, 90139 Palermo, Italy; francesco.dispenza@gmail.com
3 ENT Department, A.O.U.P. Paolo Giaccone, Via del Vespro, 129, 90127 Palermo, Italy; federicosireci@hotmail.it (F.S.); salvatore.gallina@unipa.it (S.G.)
* Correspondence: francesco.martines@unipa.it

Abstract: (1) Background: Inferior turbinates' hypertrophy is often associated with Eustachian tube dysfunction (ETD); radiofrequency turbinate reduction (RTR) may provide a long-term improvement of nasal obstruction and ETD-related symptoms. (2) Aim: The study aimed to compare ETD in atopic and non-atopic patients before and after RTR and to investigate the correlation between tympanometry and Eustachian Tube Dysfunction Questionnaire-7 (ETDQ-7). (3) Methods: Ninety-seven patients, ranging from 33 to 68 years old, were screened by skin tests and divided into atopic (G1) and non-atopic (G2). Eustachian tube function (ETF) was evaluated through tympanometry, William's test and ETDQ-7. (4) Results: A moderate to severe subjective ETDQ-7 was found in the 35.42% of G1 and in the 22.45% of G2 patients before RTR. William's test resulted normal in 141 ears (72.68%), partially impaired in 15 (7.73%), and grossly impaired in 38 (19.59%) before surgery. A grossly ETD was evidenced in the 19.59% of cases before surgery and decreased to 6.18% after surgery with a significant difference among atopic patients ($p < 0.001$). (5) Conclusion: RTR may be considered a treatment option in patients suffering from ETD and inferior turbinates' hypertrophy; RTR reduced the percentage of grossly impaired ET function ($p < 0.001$). ETDQ-7 and William's test may represent valuable tools to assess ET function before and after surgery.

Keywords: eustachian tube dysfunction (ETD); chronic nasal obstruction; turbinate hypertrophy; ETDQ-7

Citation: Martines, F.; Dispenza, F.; Sireci, F.; Gallina, S.; Salvago, P. Eustachian Tube Function Assessment after Radiofrequency Turbinate Reduction in Atopic and Non-Atopic Patients. *Int. J. Environ. Res. Public Health* **2021**, *18*, 881. https://doi.org/10.3390/ijerph18030881

Received: 20 November 2020
Accepted: 18 January 2021
Published: 20 January 2021

1. Introduction

Chronic nasal obstruction due to inferior turbinate hypertrophy is one of the most frequent ENT complaints in patients suffering from allergic and non-allergic rhinitis [1–4]; it is often associated with fullness of one or both ears, earache, tinnitus and hearing impairment. All these symptoms may be correlated to Eustachian tube dysfunction (ETD); in fact, the ET connects the middle ear with the nasopharynx, equalizing pressure between the tympanic cavity and atmospheric pressure and draining secretions from the middle ear to the nasopharynx [5–8]. ETD has an estimated prevalence ranging from 0.9% to 4% in the adult population and has been suggested as a causal factor in different ear pathologies [9–11].

Intranasal steroids, antihistamines, chromones and sympathomimetics are usually prescribed to relieve symptoms related to enlargement of the inferior turbinate, but in non-responsive patients, turbinate surgery (radiofrequency ablation, diode laser and microdebrider-assisted inferior turbinoplasty) should be suggested [12–17].

It is widely demonstrated that inferior turbinate surgery maximizes volumetric reduction of the turbinate and may provide a long-term improvement of nasal obstruction; however, data about the effect of turbinate surgery on ETD and hearing related symptoms are still scarce [18–21].

Manometric tests study ventilatory and pressure equalization abilities of the ET. One of the simplest manometric tests is tympanometry, which is the basic test in case of suspected ETD; it provides an indirect measure of ET function (ETF) by measuring middle ear pressure.

If the ET is functioning normally, middle ear pressure should be equal to atmospheric pressure, with a −20 to +20 decapascal (daPa) range in the 95% of healthy subjects [22]. However, many studies have found middle ear pressure to be slightly negative even in healthy ears and pressures from −50 to +50 daPa can be considered normal in adults [8,23,24]. This test remains particularly effective in detecting middle ear effusions [2,4,6], with a reported sensitivity and specificity of 94% and 95%, respectively [25].

Different methods were described to assess ET function using manometric tests. The easiest uses basic tympanometry equipment to look at patient-induced pressure changes in the middle ear while the patient performs a forcible 'sniff', a Valsalva or a Toynbee maneuver; however, Doyle et al. reported low percentage values of sensitivity and specificity [26].

Other common manometric tests are the Toynbee's test and automated ETF–William's test used in both perforated and intact tympanic membrane and had a reported accuracy of 81% [27]. All the other methods proposed to assess ET function (e.g., sonotubometry, tubomanometry, nasopharyngeal maneuvers, video-endoscopy, Electromyography) still have low percentage values of sensitivity (ranging from 74.2% to 87%) and specificity (ranging from 65.6% to 67%) but are also limited by the cost of equipment and the evidence that aural complaints are not always correlated to tympanometry [10–12,27–29].

For this reason, in 2012 McCoul et al. introduced the Eustachian Tube Dysfunction Questionnaire-7 (ETDQ-7) for quantitative assessment of ETD-related symptoms with an 'Ideal' of sensitivity and specificity of 100%; this validated organ-specific tool consists of seven questions with a seven-item Likert scale, with a response of 1 indicating no problem and 7 indicating a severe problem and a final score ranging from 7 to 49 points [13,14,27,30,31].

The aim of this study was to compare ETD in atopic and non-atopic patients before and after inferior turbinate surgery and to investigate the correlation between tympanometry measurements and ETDQ-7 scores.

2. Materials and Methods

This study was an observational cohort study involving 113 adult patients recruited from September 2017 to September 2019. All subjects included complained chronic nasal obstruction and at least one aural symptom (fullness or clogging of the ears, earache, tinnitus and inability to rapidly compensate middle ear pressure) which did not respond to a 3-month trial of appropriate treatment (topical corticosteroids, antihistamines, and/or sympathomimetics). Through examination by anterior rhinoscopy and nasal endoscopy, an experienced otolaryngologist confirmed the presence of bilateral inferior turbinate hypertrophy and referred the patient for radiofrequency turbinate reduction (RTR). Patients with significant nasal septum deviation, internal/external valve collapse/stenosis, chronic rhinosinusitis with or without polyposis, sinonasal and nasopharyngeal tumors, previous nasal and/or ear surgery, severe systemic disorder and severe obesity were excluded.

The study protocol was fully explained and written informed consent was obtained from each patient. Approval for this study had been obtained from the local ethical committee (Approval No. 28/06).

Skin tests were performed using 12 common perennial and seasonal allergens: *Alternaria, Aspergillus, Cladosporium, Penicillium*, ragweed, grass mix, trees mix, cockroach, dust mites, *Dermatophagoides farinae* and *Dermatophagoides pteronyssinus*, and cat and dog epithelium. Solutions of histamine and saline were used as positive and negative controls, respectively. The results were evaluated after 10 min. Wheals ≥3 mm in diameter than wheals at the site of the negative control were considered positive. The subjects with at least 1 positive skin prick test to any antigen were classified as atopic [17] and included in the group 1 (G1) while those with negative skin prick test (non-atopic) were included in the group 2 (G2).

All surgical procedures were performed by the same surgeon using a digital 130 Watt mono/bipolar unit with an operating frequency of 4 MHz and a modulation frequency of 33 kHz.

The procedures were carried out under local anesthesia with patients' eyes covered. First, the inferior turbinates were topically anesthetized using cotton strips with a mixture of lidocaine 20 mg/mL and 2–3 drops of epinephrine 0.1% in 5 to 10 mL of lidocaine. The local anesthetic (10 mg/mL lidocaine hydrochloride with 2 to 3 mL epinephrine—Lidocain) was then applied through a 24-gauge needle into the anterior and medial parts of the inferior turbinate.

The Binner bipolar needle electrode was inserted submucosally in the inferior half of the turbinate longitudinally; the first entry was in the mid anterior part of the inferior turbinate, and the additional two entries were performed medially introducing the needle as far as the posterior end of the turbinate. The parts were treated for 10 s at 35-watt (W) output power in all three areas. Because submucosally (intraturbinally) delivered radiofrequency energy in the inferior turbinate creates a focal lesion with no damage to adjacent structures (e.g., turbinate bone or mucosal surface), the patients were allowed to leave the office without medication after two hours of the completion of the treatment; they were also advised to use ibuprofen or ketoprofen for local pain or, to contact the center if needed, and do not use oral or topical steroids, antihistamines, or decongestants during the follow-up period.

All patients were evaluated two weeks before the surgery and after three months by the surgery; during the visits, the subjects filled the ETDQ-7 for quantitative assessment of ETD-related symptoms before meeting the examiner.

The ETDQ-7, a validated organ-specific tool, consists of seven questions with a seven-item Likert scale, with a response of 1 indicating no problem and 7 indicating a severe problem and a final score range from 7 to 49 points (total score <14.5 is considered normal) [30,31]. We decided to divide the total score by 7 to give an overall score ranging from 1.0 to 7.0. This equated to an ETDQ-7 mean item score of $\geq$2.1 to indicate the presence of ETD [18].

Pure tone audiometry (PTA) was performed before surgery by a trained audiologist with an Amplaid 309 audiometer in a soundproof audiometric room. Air conduction was measured on-ear TDH-49 headphones set for 250–8000 Hz; bone conduction was measured using a calibrated bone transducer for 250–4000 Hz. Mean PTA resulted as 24.15 dB HL. No cases of conductive and mixed hearing loss were detected.

The Amplaid 766tympanometer, with a probe frequency of 220 Hz and an air pressure range of −400–100 daPa with automatic recording, was used for tympanometry and to study ET function. Tympanograms were divided into the following types: type A (+99 to −99 daPa), type C (>−100 daPa) and type B (flat curve without peak identification).The tympanometry measurements considered were: ear canal volume (ECV), that is an estimation of the volume of air medial to the probe, which includes the volume between the probe tip and the tympanic membrane; the tympanometry peak pressure (TTP), corresponding to ear canal pressure at which the peak of the C tympanograms occurs; the static compliance (SC), that describes the greatest amount of acoustic energy absorbed by the middle ear system.

ET function was tested through William's test in which the impedance audiometer is programmed to measure the middle ear pressure in 3 consecutive conditions: at the start of the test (resting pressure—Peak 1), after the patient swallows (with the nose and mouth closed—Peak 2), and finally after performing Valsalva (Peak 3). Normally, the ambient (i.e., resting) middle ear pressure should be at or near the atmospheric air pressure (i.e., approx. 0 mm of water pressure); the tympanic cavity pressure should become negative on swallowing and positive on performing Valsalva. The findings of the ETF-William's test for the different groups of patients were classified into 3 categories: (1) Perfectly normal function, (2) Partially impaired function and (3) Grossly impaired function. To evaluate the functionality of the Eustachian tube, we used the Peak 1-Peak 2 > 20 daPa (after swallow) and Peak 3-Peak 1 > 20 daPa (after Valsalva) criteria. If the ambient middle ear pressure becomes negative on swallowing but does not become positive on Valsalva or vice versa the function is considered to be partially impaired. If the middle ear pressure does not change either after swallowing and after Valsalva maneuver, the ET function is considered grossly impaired.

Statistical analysis was conducted performing χ^2 test, odds ratio (OR), correlation analysis (r) and *t* test, following usual conditions of application. Significance was set at 0.05.

3. Results

Out of 113 patients, 5 subjects were excluded because of severe systemic disorders diagnosed during the admission analysis; a total of 108 patients underwent surgery but finally 97 (mean age = 48.2 ± 9.1) were included in the study because 11 patients were treated with antihistamines associated to selective leukotriene receptor antagonists during follow-up period. The age range was 33–68 years with a mean age of 48.22 ± 9.04.

Based on skin tests, they were classified in 2 groups: G1 (atopic) and G2 (non-atopic), consisting of 48 and 49 subjects, respectively.

Among G1 group, 22 patients (45.83%) had positive skin tests for both inhalant and food allergens; 9 patients (18.75%) had a positive test only for food allergy; 17 (35.42%) patients had an allergy only against inhalant allergens.

Sixty-nine patients had a preoperative ETDQ-7 total score of 1.0 to 1.9 (without subjective dysfunction), 12 presented a 2.0 to 2.9 score (moderate ETD) and 16 had a score >3.0 (severe ETD) with a range of 3.0–3.7 (Figure 1). The distribution of ETDQ-7 total score among the groups evidenced a ETD in 35.42% of atopic (8 subjects with a score of 2.0 to 2.9 while 9 with a score >3.0) and in the 22.45% of non-atopic patients (4 subjects with a score of 2.0 to 2.9 and 7 with a score >3.0) preoperatively. The percentage of patients with a normal preoperative ETDQ-7 resulted respectively 64.58% and 77.55% for G1 and G2 groups without a statistically significant difference ($p = 0.319$) (Figure 1).

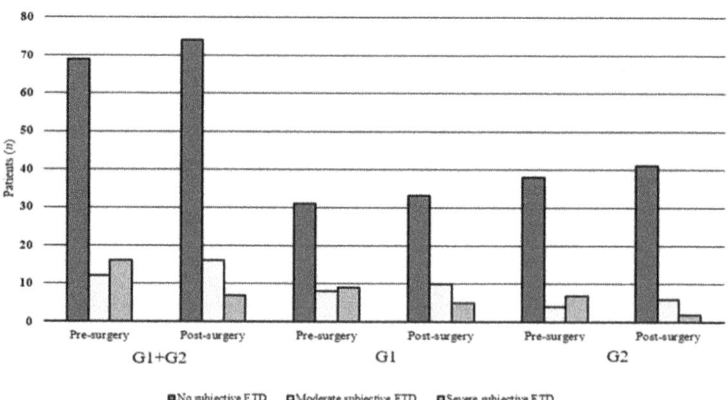

Figure 1. Eustachian Tube Dysfunction Questionnaire-7 (ETDQ-7) among atopic and non-atopic patients.

The study of the mean values of single ETDQ-7 items showed higher mean scores in atopic respect to non-atopic patients; a significant difference was found among atopic subjects between pre and post-surgery only in case of the following question: 'A feeling that your ears are clogged?' ($p = 0.049$); no difference was found regarding the other items ($p > 0.05$) (Table 1).

Table 1. Eustachian Tube Dysfunction Questionnaire-7 (ETDQ-7) scores before and after surgery.

ETDQ-7	Atopic + Non-Atopic G1 + G2			Atopic G1			Non-Atopic G2		
Items	**Pre-Surgery Mean (SD)**	**Post-Surgery Mean (SD)**	**T-Test (*p*-Value)**	**Pre-Surgery Mean (SD)**	**Post-Surgery Mean (SD)**	**T-Test (*p*-Value)**	**Pre-Surgery Mean (SD)**	**Post-Surgery Mean (SD)**	**T-Test (*p*-Value)**
Pressure in the ears?	1.75 ± 0.81	1.55 ± 0.79	0.09	1.87 ± 0.87	1.72 ± 0.89	0.41	1.63 ± 0.75	1.38 ± 0.63	0.08
Pain in the ears?	1.87 ± 1.37	1.64 ± 1.10	0.21	2.12 ± 1.52	1.89 ± 1.29	0.43	1.63 ± 1.18	1.41 ± 0.83	0.28
A feeling that your ears are clogged or "under water"?	1.94 ± 0.96	1.69 ± 0.89	0.05	2.25 ± 0.88	1.87 ± 0.95	0.04	1.65 ± 0.95	1.51 ± 0.79	0.42
Ear symptoms when you have a cold or sinusitis?	2.21 ± 1.20	1.97 ± 1.01	0.14	2.41 ± 1.25	2.23 ± 1.03	0.42	2.02 ± 1.14	1.73 ± 0.93	0.18
Crackling or popping sounds in the ears?	1.61 ± 0.99	1.47 ± 0.84	0.31	1.73 ± 1.04	1.58 ± 0.91	0.47	1.49 ± 0.93	1.36 ± 0.75	0.48
Ringing in the ears?	1.27 ± 0.55	1.26 ± 0.54	0.89	1.29 ± 0.50	1.27 ± 0.49	0.84	1.26+0.60	1.26 ± 0.60	1
A feeling that your hearing is muffled?	1.78 ± 0.89	1.64 ± 0.75	0.25	1.83 ± 0.86	1.68 ± 0.77	0.38	1.73 ± 0.93	1.61 ± 0.73	0.47
Mean Total Score (Min–Max)	12.46 ± 5.77 (8–26)	11.26 ± 5.04 (7–26)	0.12	13.52 ± 6.11 (8–26)	12.27 ± 5.61 (7–26)	0.29	11.42 ± 5.28 (8–25)	10.28 ± 4.25 (7–21)	0.24
Mean Item Score (Min–Max)	1.8 ± 0.8 (1.1–3.7)	1.6 ± 0.7 (1.0–3.7)	0.12	1.9 ± 0.9 (1.1–3.7)	1.7 ± 0.8 (1.0–3.7)	0.29	1.6 ± 0.7 (1.1–3.5)	1.5 ± 0.6 (1.0–3.0)	0.24

Among tympanometry measurements TTP ranges from −62 to +39 daPa and from −60 to +41 daPa respectively for G1 and G2 evidencing a type "A" tympanogram for all patients ($p > 0.05$).

The ECV and the SC mean values for the total ears resulted in being 1.73 ± 0.69 cc and 1.18 ± 0.37 cc with a significant difference between the groups (G1: ECVmean value = 1.85 ± 0.62 cc; SC mean value 1.30 ± 0.28 cc; G2: ECV mean value = 1.60 ± 0.72 cc; SC mean value: 1.06 ± 0.41 cc; $p < 0.01$) (Table 2). In the 42.78% of cases for ECV and in the 6.63% of patients for SC the results were out of normal range with a higher percentage inside the atopic group ($p < 0.01$).

Out of 194 ears examined, the William's test resulted normal in 141 cases (72.68%), partially impaired in 15 (7.73%) and grossly impaired in 38 (19.59%). The distribution among the groups evidenced a significant difference with an ET normal function in the 79.59% of G2 group with respect to the 65.62% of G1 group ($p = 0.003$) (Table 3).

In the 81.25% of patients with a severe ETDQ-7 score a grossly impaired ET function was evidenced while 94.96% of subjects with a normal ETDQ-7 presented a normal ET function ($p < 0.001$) with a correlation index r = 0.99 (Table 3).

The ETDQ-7 filled by patients three months after surgery showed an improvement with a 76.29% of normal ET function (74 cases, 33 G1 and 41 G2), a 16.49% with a moderate ETD (16 cases, 10 G1 and 6 G2) and a 7.22% (5 G1 subjects and 2 G2) reporting severe ETD ($p = 0.014$). Also, the distribution of the mean values of single ETDQ-7 items showed an improvement between pre- and post-surgery related to G1, G2 and total cohort but without any significant difference ($p = 0.1$) (Table 1).

At follow up examination we observed an improvement of ECV (1.50 ± 0.46 cc) and SC (1.17 ± 0.28 cc) in most of cases; in fact, the percentage values out of normal range decreased to 26.80% and to 4.12% for ECV and SC, respectively. A significant difference was evidenced for ECV ($p < 0.0001$). The differences in tympanometry measurement between G1 and G2 groups were also confirmed three months after surgery (G1: ECV mean value = 1.60 ± 0.49 cc; SC mean value 1.23 ± 0.29 cc; G2: ECV mean value = 1.39 ± 0.38 cc; SC mean value: 1.11 ± 0.25 cc; $p = 0.002$) (Table 2).

William's test results changed significantly during follow up examination; we noted a reduction of 'grossly ET impaired function' that decreased from 19.59% to 6.18% respect to a higher percentage of 'partial impaired function' that increased from 7.73% to 21.65% ($p < 0.001$). This statistical difference was more evident inside G1 population where 'grossly ETD' decreased from 29.17% to 8.33% and 'partial impaired function' percentage changed from 5.21% to 26.04% ($p < 0.001$).

As seen before the surgery, also three months after surgery the distribution of William's test and ETDQ-7 scores in the cohort resulted as being statistically significant ($p < 0.001$); in fact, the 84.45% of subjects with a normal ETF presented a normal ETDQ-7 with a correlation index r = 0.87.

Table 2. William's test and tympanometry measurements before and after surgery.

William's Test	**Atopic + Non-Atopic G1 + G2 Ears (194)**			**Atopic G1 Ears (96)**			**Non-Atopic G2 Ears (98)**		
	Pre-surgery ***n*** (%)	Post-surgery ***n*** (%)	χ^2 (*p*-value)	Pre-surgery ***n*** (%)	Post-surgery ***n*** (%)	χ^2 (*p*-value)	Pre-surgery ***n*** (%)	Post-surgery ***n*** (%)	χ^2 (*p*-value)
Normal	141 (72.68)	140 (72.16)		63 (65.62)	63 (65.62)		78 (79.59)	77 (78.57)	
Partially Impaired	15 (7.73)	42 (21.65)	(0.0001)	5 (5.21)	25 (26.04)	(0.0001)	10 (10.20)	17 (17.35)	
Toymbe	7	15		2	10		5	5	
Valsalva	8	27		3	15		5	12	0.1
Grossly Impaired	38 (19.59)	12 (6.18)		28 (29.17)	8 (8.33)		10 (10.20)	4 (4.08)	
Tympanometry measurements			T-Test (*p*-value)			T-Test (*p*-value)			T-Test (*p*-value)
ECV (cc) Mean ± SD	1.73 ± 0.69	1.50 ± 0.46	(0.0001)	1.85 ± 0.62	1.60 ± 0.49	(0.002)	1.60 ± 0.72	1.39 ± 0.38	(0.01)
SC (cc) Mean ± SD	1.18 ± 0.37	1.17 ± 0.28	(0.72)	1.30 ± 0.28	1.23 ± 0.29	(0.10)	1.06 ± 0.41	1.11 ± 0.25	(0.31)

ECV: ear canal volume; SC: static compliance.

Table 3. ETDQ-7 score and William's test assessment before and after surgery.

ETDQ-7	William's Test					
		Pre-Surgery			Post-Surgery	
	Normally	Partially Impaired	Grossly Impaired	Normally	Partially Impaired	Grossly Impaired
1.0–1.9 No subjective ETD	132	6	0	125	23	0
2.0–2.9 Moderately subjective ETD	7	5	12	13	14	5
>3.0 Severe subjective ETD	2	4	26	2	5	7
TOTAL	141	15	38	140	42	12

Correlation index (r)—pre-surgery: r = 0.99; post-surgery: r = 0.87. ETDQ-7 pre- and post-surgery: $p = 0.014$. William's Test pre- and post-surgery: $p < 0.001$. ETD: Eustachian tube dysfunction.

4. Discussion

The relationship between the middle ear ventilation, ET, and nasal cavities has been the subject of numerous studies [2,3,5,9,11,15,25,32–35]; Low and Willatt in 1993 evaluating the relationship between middle ear pressure and a deviated nasal septum observed an improvement of TTP in both ears after septal surgery [32]; the authors postulated that postnasal airflow turbulence associated with a deviated nasal septum could lead to ETD. It could be explained because several patients, either atopic or non-atopic, usually complain to suffer from middle ear ventilation problems concomitant with nasal obstructive pathologies [1,2,6,14,28,32–34]. Certainly, all authors agreed that ETD is thought to be an important cause of middle ear disturbances [1–8,25,32,35]. Despite an initial pharmacological treatment, patients with inferior turbinates' hypertrophy and nasal obstruction often undergo turbinate radiofrequency ablation [12–16,19]. This surgery usually provides good long-term results but the changes on ET function and improvement of the ear related symptoms are usually not investigated or partially evaluated using only tympanometry [18–21,32,33].

In 2012 McCoul et al. introduced the Eustachian Tube Dysfunction Questionnaire-7 (ETDQ-7), now universally recognized as a good tool to identify people with subjective and objective ETD [29,30].

The percentage of abnormal tympanometry measurements and altered William's test among patients suffering from chronic nasal obstruction varies with a range from 12.5% to 58% [36,37]. In line with Lazo-Saenz et al. (2005), who reported either doubled percentages of ETD and higher ECV and SC mean values in atopic cohort with respect to control group, we observed a grossly ETD in the 29.17% of G1 group with respect to the 10.20% of G2 cohort, and a higher value of ECV and SC out of normal range in G1 respect to G2 (56% vs. 29.59% for ECV and 10.41% vs. 3.06% for SC) [38].

Some authors demonstrated that surgery improved tympanometry and ET function, while others reported a not significant improvement in ET function postoperatively [32–34,36,37]. In line with Salvinelli et al. who demonstrated an improved ET function after nasal surgery, we observed a significant reduction of 'grossly ETD'; in fact, out of 38 ears with grossly ETD (28 ears G1 and 10 ears G2), only the 6.18% (12 ears: 8 belonging to G1 and 4 to G2) did not show any improvement after surgery ($p < 0.001$) [35]. Moreover, among tympanometry measures, there was a significant normalization of ECV average during the follow-up ($p < 0.001$).

In a study conducted by Doğan et al., 80 patients with six different septum types and consequent nasal obstruction underwent tympanometry measurement and Automated ETF (based on William's test) using Impedance Audiometry AC40 before and 6 months after surgery [39]. The author considered the 'Peak 1-Peak 2 > 10 daPa or Pmax-Pmin > 15 daPa' criteria to study ETF; all measurements improved 6 months after surgery and when the obstruction was localized in the inferior meatus and in the floor of nasal cavity the improvement resulted significant. Using the same criteria, we obtained similar results after RTR; comparing pre- and post-surgery Peak 1-Peak 2 and Peak 3-Peak1 mean values, we demonstrated how volumetric reduction of the turbinates improved significantly the William's test results ($p < 0.001$).

Therefore, RTR could be considered an option for those subjects who complain ETD associated to chronic nasal obstruction due to inferior turbinate hypertrophy with or without atopy.

Even if ETDQ-7 was validated in 2012, only from 2017 it was used by Harju et al. to study the effect of inferior turbinate surgery on ETD-related symptoms [18,30,31]. The authors showed a statistically significant improvement in all subjects, either after real surgeries that were limited to the anterior part ($p = 0.03$ for radiofrequency ablation and $p = 0.006$ for microdebrider-assisted inferior turbinoplasty) and after sham surgery ($p = 0.04$). For this reason, Harju et al. concluded that reduction of the anterior half of the inferior turbinate, as the only procedure to treat the ear symptoms, is useless because the operation does not involve the posterior part of the turbinate, which is close to the ET orifice.

In our cohort, the ETDQ-7 questionnaire filled before surgery resulted reliable; in fact, in the 81.25% of a severe ETDQ-7 score it was evidenced a grossly impaired ET function at William's test and the 94.96% of subjects with a normal ETDQ-7 had a normal ET function ($p < 0.001$). Compared to Harju et al. who observed before surgery a normal ETDQ-7 in 45% of patients, we had higher normal values (71.13%) especially among non-atopic patients (77.55%) that usually had less hearing-related complains. This is reasonable because for three months before surgery all patients underwent to oral and/or topical nasal therapy that could have improved aural symptoms present 'ab origine'. The ETDQ-7 reliability was also confirmed after surgery where a normal ETDQ-7 was found in the 84.45% of subjects with a normal ET function ($r = 0.87$; $p < 0.001$).

The ETDQ-7 mean value decreased from 1.8 before surgery to 1.6 at follow up ($p = 0.12$); it overlaps to the normal ETF percentage at William's test between pre (72.68%) and post-surgery (72.16%) that unchanged. However, it should not be interpreted as a failure to treat ETD with RTR; in fact, grossly impaired ETF has dropped ($p = 0.014$) and the ETF (Peak 1-Peak 2: $r = 0.89$; Peak 3-Peak 1) improved significantly both in atopic and non-atopic patients.

This study presents the following limitations: to begin with, all patients were assessed only once at three months after surgery, thus we could not be able to rule out fluctuations of ETF during the 3-month period or in the long-term. Secondly, the study sample is relatively small, and results are difficult to compare because data regarding RTR and ETD are still scarce.

5. Conclusions

The present study confirmed the close relationship between aural symptoms and chronic nasal obstruction due to inferior turbinate hypertrophy. In particular, atopic patients presented a higher prevalence of ETD (34.38% vs. 20.40%, $p = 0.044$) and a higher ETDQ-7 (35.42% vs. 22.45%, $p = 0.15$) score respect to non-atopic subjects.

RTR may be considered an option in chronic ETD with inferior turbinates' hypertrophy which does not respond to medical therapy; in fact, as highlighted by our results, surgical treatment reduced the percentage of grossly impaired ETF ($p < 0.001$), normalized the ECV average ($p < 0.001$) and increased ETF both in atopic and non-atopic patients.

With a correlation index of $r = 0.99$ before and of 0.87 after surgery ($p < 0.001$), ETDQ-7 and William's test may represent valuable tools to assess both subjectively and objectively candidates for RTR.

Author Contributions: Conceptualization, F.M. and P.S.; methodology, F.S.; software, F.D.; validation, F.M., P.S. and F.D.; formal analysis, F.M.; investigation, S.G.; resources, S.G.; data curation, F.S.; writing—original draft preparation, F.M.; writing—review and editing, P.S.; visualization, F.S.; supervision, F.M.; project administration, F.M.; funding acquisition, F.D. All authors have read and agreed to the published version of the manuscript.

Funding: This research received no external funding.

Institutional Review Board Statement: The study was conducted according to the guidelines of the Declaration of Helsinki, and approved by the Ethics Committee of the AOUP "P. Giaccone" (protocol code 28/06, date of approval: 15/07/17).

Informed Consent Statement: Informed consent was obtained from all subjects involved in the study.

Data Availability Statement: The data are not publicly available due to privacy restrictions.

Conflicts of Interest: The authors reported no potential conflict of interest.

References

1. Pelikan, Z. Chronic otitis media (secretory) and nasal allergy. *Scr. Med.* **2006**, *79*, 177–178.
2. Bentivegna, D.; Salvago, P.; Agrifoglio, M.; Ballacchino, A.; Ferrara, S.; Mucia, M.; Sireci, F.; Martines, F. The linkage between Upper Respiratory Tract Infections and Otitis Media: Evidence of the 'United airways concept'. *Acta Med. Medit.* **2012**, *28*, 287–290.
3. Tewfik, T.L.; Mazer, B. The links between allergy and otitis media with effusion. *Curr. Opin. Otolaryngol. Head Neck Surg.* **2006**, *14*, 187–190. [CrossRef] [PubMed]
4. Martines, F.; Salvago, P.; Ferrara, S.; Messina, G.; Mucia, M.; Plescia, F.; Sireci, F. Factors influencing the development of otitis media among Sicilian children affected by upper respiratory tract infections. *Braz. J. Otorhinolaryngol.* **2016**, *82*, 215–222. [CrossRef] [PubMed]
5. Casselbrant, M.L.; Mandel, E.M. Epidemiology. In *Evidence-Based Otitis Media*, 2nd ed.; Rosenfeld, R.M., Bluestone, C.D., Eds.; BC Decker: Hamilton, ON, Canada, 2000; pp. 147–162.
6. Martines, F.; Bentivegna, D. Audiological investigation of Otitis Media in Children with Atopy. *Curr. Allergy Asthma Rep.* **2011**, *11*, 513–520. [CrossRef] [PubMed]
7. Mucia, M.; Salvago, P.; Brancato, A.; Cannizzaro, C.; Cannizzaro, E.; Gallina, S.; Ferrara, S.; La Mattina, E.; Mulè, A.; Plescia, F.; et al. Upper respiratory tract infections in children: From case history to management. *Acta Med. Medit.* **2015**, *31*, 419–424.
8. Sade, J.; Ar, A. Middle ear and auditory tube: Middle ear clearance, gas exchange, and pressure regulation. *Otolaryngol. Head Neck Surg.* **1997**, *116*, 499–524. [CrossRef]
9. Browning, G.G.; Gatehouse, S. The prevalence of middle ear disease in the adult British population. *Clin. Otolaryngol. Allied Sci.* **1992**, *17*, 317–321. [CrossRef]
10. Ferrara, S.; Di Marzo, M.; Martines, F.; Ferrara, P. Medical and surgical update on "atelectasic-Adhesive-Tympanosclerotic" otitis media. *Otorinolaringologia* **2011**, *61*, 11–17.
11. Bluestone, C.D. *Eustachian Tube. Structure, Function, Role in Otitis Media*; BC Decker Inc.: Hamilton, OH, USA; London, UK, 2005.
12. Hillas, J.; Booth, R.J.; Somerfield, S.; Morton, R.; Avery, J.; Wilson, J.D. A comparative trial of intranasal beclomethasone dipionate and sodium chromoglycate in patients with chronic perennial rhinitis. *Clin. Allergy* **1980**, *10*, 253–258. [CrossRef]
13. Kirkegaard, J.; Mygind, N.; Molgaard, F.; Grahne, B.; Holopainen, E.; Malmberget, H.; Brøondbo, K.; Røjne, T. Ordinary and high-dose ipratropium in perennial nonallergic rhinitis. *J. Allergy Clin. Immunol.* **1987**, *79*, 585–590. [CrossRef]
14. Krause, H. Antihistamines and decongestants. *Otolaryngol. Head Neck Surg.* **1992**, *6*, 835–840. [CrossRef] [PubMed]
15. Berger, G.; Gass, S.; Ophir, D. The histopathology of the hypertrophic inferior turbinate. *Arch. Otolaryngol. Head Neck Surg.* **2006**, *132*, 588–594. [CrossRef] [PubMed]
16. Willatt, D. The evidence for reducing inferior turbinates. *Rhinology* **2009**, *47*, 227–236. [CrossRef] [PubMed]
17. Fokkens, W.J.; Lund, V.J.; Hopkins, C.; Hellings, P.W.; Kern, R.; Reitsma, S.; Toppila-Salmi, S.; Bernal-Sprekelsen, M.; Mullol, J.; Alobid, I.; et al. European Position Paper on Rhinosinusitis and Nasal Polyps 2020. *Rhinology* **2020**, *58* (Suppl. 29), 1–464. [CrossRef]
18. Harju, T.; Kivekäs, I.; Numminen, J.; Rautiainen, M. The effect of inferior turbinate surgery on ear symptoms. *Laryngoscope* **2018**, *128*, 568–572. [CrossRef]
19. Li, K.K.; Powell, N.B.; Riley, R.W.; Troell, R.J.; Guilleminault, C. Radiofrequency volumetric tissue reduction for treatment of turbinate hypertrophy: A pilot study. *Otolaryngol. Head Neck Surg.* **1998**, *119*, 569–573. [CrossRef]
20. Elwany, S.; Gaimaee, R.; Fattah, H.A. Radiofrequency bipolar submucosal diathermy of the inferior turbinates. *Am. J. Rhinol.* **1999**, *13*, 145–149. [CrossRef]
21. Coste, A.; Yona, L.; Blumen, M.; Louis, B.; Zerah, F.; Rugina, M.; Peynègre, R.; Harf, A.; Escudier, E. Radiofrequency is a safe and effective treatment of turbinate hypertrophy. *Laryngoscope* **2001**, *111*, 894–899. [CrossRef]
22. British Society of Audiology. *Tympanometry: Recommended Procedure*; British Society of Audiology: Reading, UK, 2013.

23. Seifert, M.W.; Seidemann, M.F.; Givens, G.D. An examination of variables involved in tympanometric assessment of Eustachian tube function in adults. *J. Speech Hear. Disord.* **1979**, *44*, 388–396. [CrossRef]
24. Shanks, J.; Shelton, C. Basic principles and clinical applications of tympanometry. *Otolaryngol. Clin. N. Am.* **1991**, *24*, 299–328. [CrossRef]
25. Shekelle, P.; Takata, G.; Chan, L.; Mangione-Smith, R.; Corley, P.M.; Morphew, T.; Morton, S. Diagnosis, Natural History, and Late Effects of Otitis Media with Effusion. *Evid. Rep./Technol. Assess.* **2002**, *55*, 1–5.
26. Doyle, W.J.; Swarts, J.D.; Banks, J.; Casselbrant, M.L.; Mandel, E.M.; Alper, C.M. Sensitivity andspecificity of eustachian tube function tests in adults. *JAMA Otolaryngol. Head Neck Surg.* **2013**, *139*, 719–727. [CrossRef] [PubMed]
27. Smith, M.E.; Tysome, J.R. Tests of Eustachian tube function: A review. *Clin. Otolaryngol.* **2015**, *40*, 300–311. [CrossRef]
28. Liu, P.; Su, K.; Zhu, B.; Wu, Y.; Shi, H.; Yin, S. Detection of Eustachian tube openings by tubomanometry in adult otitis media with effusion. *Eur. Arch. Otorhinolaryngol.* **2016**, *273*, 3109–3115. [CrossRef]
29. Van der Avoort, S.J.; van Heerbeek, N.; Zielhuis, G.A.; Cremers, C.W. Sonotubometry: Eustachian tube ventilatory function test: A state-of-the-art review. *Otol. Neurotol.* **2005**, *26*, 538–543. [CrossRef]
30. McCoul, E.D.; Anand, V.K.; Christos, P.J. Validating the clinical assessment of eustachian tube dysfunction: The Eustachian Tube Dysfunction Questionnaire (ETDQ-7). *Laryngoscope* **2012**, *122*, 1137–1141. [CrossRef]
31. Teixeira, M.S.; Swarts, J.D.; Alper, C.M. Accuracy of the ETDQ-7 Questionnaire for Identifying Persons with Eustachian Tube Dysfunction. *Otolaryngol. Head Neck Surg.* **2018**, *158*, 83–89. [CrossRef]
32. Low, W.K.; Willatt, D.J. The relationship between middle ear pressure and deviated nasal septum. *Clin. Otolaryngol. Allied Sci.* **1993**, *8*, 308–310. [CrossRef]
33. Nanda, M.S.; Kaur, M.; Bhatia, S. Impact of septoplasty on hearing and middle ear function. *Int. J. Res. Med. Sci.* **2018**, *6*, 135–138. [CrossRef]
34. Şereflican, M.; Yurttaş, V.; Oral, M.; Yılmaz, B.; Dağlı, M. Is middle ear pressure affected by nasal packings after septoplasty? *J. Int. Adv. Otol.* **2015**, *11*, 63–65. [CrossRef] [PubMed]
35. Kaya, M.; Dağlı, E.; Kırat, S. Does Nasal Septal Deviation Affect the Eustachian Tube Function and Middle Ear Ventilation? *Turk. Arch. Otorhinolaryngol.* **2018**, *56*, 102–105. [CrossRef] [PubMed]
36. Salvinelli, F.; Casale, M.; Greco, F.; D'Ascanio, L.; Petitti, T.; Di Peco, V. Nasal surgery and Eustachian tube function: Effects on middle ear ventilation. *Clin. Otolaryngol.* **2005**, *30*, 409–413. [CrossRef] [PubMed]
37. Akyildiz, M.Y.; Özmen, Ö.A.; Demir, U.L.; Kasapoğlu, F.; Coşkun, H.H.; Basut, O.I.; Siğirli, D. Impact of septoplasty on Eustachian tube functions. *J. Craniofac. Surg.* **2017**, *28*, 1929–1932. [CrossRef] [PubMed]
38. Lazo-Sáenz, J.G.; Galván-Aguilera, A.A.; Martínez-Ordaz, V.A.; Velasco-Rodríguez, V.M.; Nieves-Rentería, A.; Rincón-Castañeda, C. Eustachian tube dysfunction in allergic rhinitis. *Otolaryngol. Head Neck Surg.* **2005**, *32*, 626–629.
39. Doğan, R. The Effect of Types of Nasal Septum Deviation on the Eustachian Tube Function. *Bezmialem Sci.* **2019**, *7*, 33–37. [CrossRef]

Article

Dextromethorphan Attenuates Sensorineural Hearing Loss in an Animal Model and Population-Based Cohort Study

Hsin-Chien Chen [1,*], Chih-Hung Wang [1,2], Wu-Chien Chien [3,4], Chi-Hsiang Chung [3,4], Cheng-Ping Shih [1], Yi-Chun Lin [1,2], I-Hsun Li [5,6], Yuan-Yung Lin [1,2] and Chao-Yin Kuo [1]

1 Department of Otolaryngology-Head and Neck Surgery, Tri-Service General Hospital, National Defense Medical Center, Taipei 114, Taiwan; chw@ms3.hinet.net (C.-H.W.); zhengping_shi@yahoo.com.tw (C.-P.S.); lyc_1023@yahoo.com.tw (Y.-C.L.); yking1109@gmail.com (Y.-Y.L.); chefsketchup@hotmail.com (C.-Y.K.)
2 Graduate Institute of Medical Sciences, National Defense Medical Center, Taipei 114, Taiwan
3 School of Public Health, National Defense Medical Center, Taipei 114, Taiwan; chienwu@mail.ndmctsgh.edu.tw (W.-C.C.); g694810042@gmail.com (C.-H.C.)
4 Department of Medical Research, Tri-Service General Hospital, National Defense Medical Center, Taipei 114, Taiwan
5 Department of Pharmacy Practice, Tri-Service General Hospital, National Defense Medical Center, Taipei 114, Taiwan; lhs01077@gmail.com
6 School of Pharmacy, National Defense Medical Center, Taipei 114, Taiwan
* Correspondence: acolufreia@yahoo.com.tw; Tel.: +886-2-8792-7192; Fax: +886-2-8792-7193

Received: 31 July 2020; Accepted: 28 August 2020; Published: 31 August 2020

Abstract: The effect of dextromethorphan (DXM) use in sensorineural hearing loss (SNHL) has not been fully examined. We conducted an animal model and nationwide retrospective matched-cohort study to explore the association between DXM use and SNHL. Eight-week-old CBA/CaJ hearing loss was induced by a white noise 118 dB sound pressure level for 3 h. DXM (30 mg/kg) was administered intraperitoneally for 5 days and boost once round window DXM socking. In population-based study, we examined the medical records over 40 years old in Taiwan's National Health Insurance Research Database between 2000 and 2015 to establish retrospective matched-cohort to explore the correlation between DXM use and SNHL. Using click auditory brainstem response (ABR), hearing threshold was measured as 48.6 ± 2.9 dB in control mice compared with 42.6 ± 7.0 dB in DXM mice, which differed significantly ($p = 0.002$) on day 60 after noise exposure with a larger ABR wave I amplitude in DXM mice. In human study, we used a Cox regression hazard model to indicate that a significantly lower percentage individuals developed SNHL compared with and without DXM use (0.44%, 175/39,895 vs. 1.05%, 1675/159,580, $p < 0.001$). After adjustment for age and other variables [adjusted hazard ratio: 0.725 (95% confidence interval: 0.624–0.803, $p < 0.001$)], this study also demonstrated that DXM use appeared to reduce the risk of developing SNHL. This animal study demonstrated that DXM significantly attenuated noise-induced hearing loss. In human study, DXM use may have a protective effect against SNHL.

Keywords: dextromethorphan; noise; hearing loss; cochlea; synapse

1. Introduction

Hearing loss is a growing and alarmingly high burden in the world reporting from the Global Burden of Disease Studies and the third leading cause of years lived with disability [1,2]. Sensorineural hearing loss (SNHL) is primarily due to the degeneration of hair cells and spiral ganglion neurons in the cochlea resulting from acute and/or chronic events of extrinsic (e.g., ototoxic drugs, noise) and intrinsic causes (e.g., aging, genetic factors, congenital risk factors) [3,4]. The common causes of genetic, age-related, noise-induced, and drug-induced hearing loss display intriguing similarities in terms of particular cellular responses of the cochlear sensory cells comprising potential involvement of impaired mitochondrial function, ischemia, oxidative stress with reactive oxygen species, inflammation, apoptosis, autophagy, and/or necrosis [3].

Recently, a recent mechanism of SNHL is reported to be related with cochlear or auditory synaptopathy, which is caused by damage or loss to the synapses between inner hair cell (IHC) and spiral ganglion neuron (SGN) and causing deafferentation [5–8]. Cochlear synpatopathy is also involved in genetic, noise, ototoxicity and age-related hearing loss [7–10]. This mechanism may contribute to glutamate excitotoxicity involving *N*-methyl-D-aspartate (NMDA) receptor activation and related auditory nerve excitation [9,11,12]. NMDA receptor inhibition has been proposed as a pharmacologic approach for the treatment of synaptic hearing loss [13–15].

NMDA antagonists include ketamine, esketamine, dextromethorphan (DXM), phencyclidine, and dizocilpine [16,17]. DXM, an uncompetitive and low-affinity NMDA receptor antagonist, has been widely used as a nonopioid, nonnarcotic, and over-the-counter antitussive for over 50 years [16]. DXM has demonstrated considerable neuroprotective properties in numerous in vitro and in vivo models of central nervous system injury and neurodegenerative diseases [16,18]. DXM may exert its neuroprotective effects through multiple actions to inhibit glutamate neurotoxicity, inflammatory pathways, oxidative damage, calcium imbalances, and apoptosis with extremely similarity of pathophysiologies in SNHL and may play a role in sudden SNHL [3,16,19].

DXM has been reported to reduce neuronal damage or degeneration, cortical infarct volume and improve neurological functions in numerous animal models of stroke and traumatic brain injury (TBI) [18,20]. In 2010, the U.S. Food & Drug Administration approved the use of DXM in combination with quinidine for the treatment of pseudobulbar affect characterized by sudden and involuntary episodes of crying, laughing, or other emotional displays secondary to a neurological disease or brain injury, such as amyotrophic lateral sclerosis, stroke, Alzheimer's disease, multiple sclerosis, Parkinson's disease, and TBI [16]. To date, except our previous published paper, there is sparse study to explore the association between DXM use and preventing SNHL in inner ear of mouse model and human study [21]. Herein, we explore whether DXM, an NMDA antagonist, exhibits any potential effect against SNHL in both animal models and population-based study.

2. Materials and Methods

2.1. Animals and Noise Exposure

All experiments were approved by the Institutional Animal Care and Use Committee of the National Defense Medical Center (Taipei, Taiwan, approval number IACUC-16-082). The animal care complied with the institutional guidelines and regulations. The schedule of the experiment is illustrated in Figure 1A. Male eight-week old CBA/CaJ mice are randomized into control and DXM group. In total, 22 mice were operated in the DXM group and 18 in the control group. The mice were anesthetized, placed in a soundproof booth with a loudspeaker (V12 HP, Tannoy, United Kingdom) mounted above the center of the cage, and both ears were exposed to white noise at a sound pressure level (SPL) of 118 dB for 3 h. The sound intensity inside the chamber was tested using a sound level meter to ensure minimal deviations of sound intensity.

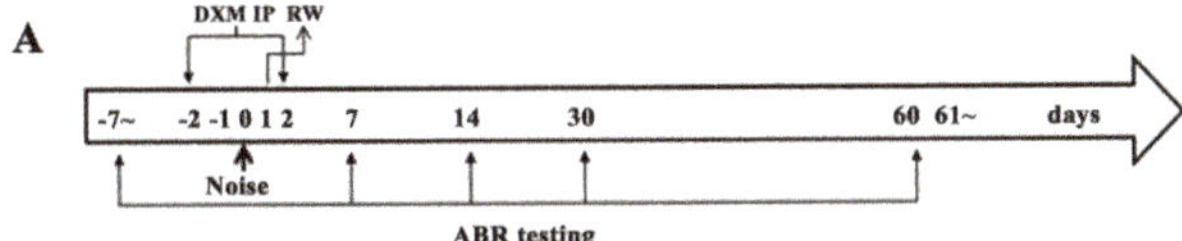

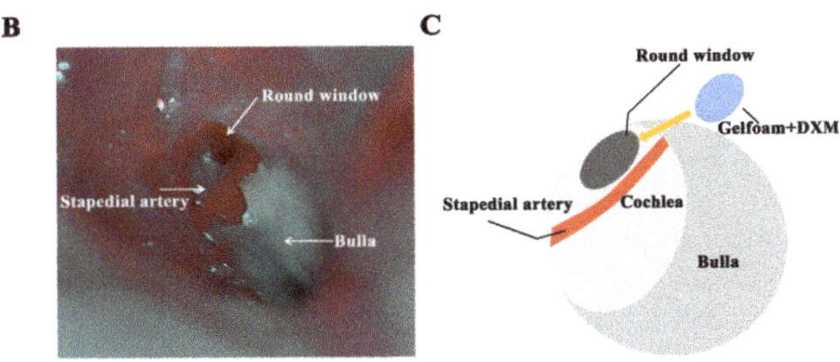

Figure 1. (**A**) Experimental schedule. Mice were exposed to noise on day 0. The auditory brainstem response (ABR) test was performed on days −7, 7, 14, 30, and 60. Dextromethorphan (DXM) was injected intraperitoneally (IP) on days −2 to 2 and boosted on day 1 through the round window approach. (**B**,**C**) Surgical approach for DXM round-window application.

2.2. DXM Application

DXM hydrobromide was purchased from Sigma-Aldrich (St. Louis, MO, USA). Each dose (30 mg/kg) was injected intraperitoneally once/per day for 5 contiguous days starting 2 days before noise exposure. The next day after noise exposure, the animals' unilateral round windows were surgically exposed to boost DXM soaking to enhance the possible therapeutic effect and future application of therapy clinically by intratympanic injection. The control group was administered phosphate-buffered saline (PBS) during the same surgical procedure (sham). The animals were anesthetized with ketamine (100 mg/kg, intraperitoneally) and xylazine (10 mg/kg, intraperitoneally). Half the initial dose of anesthesia was administered on the reappearance of blinking or the withdrawal reflex in the animals. A posteroinferior skin incision was made in the retroauricular area behind the right ear. The underlying muscles and facial nerve were separated by blunt dissection to expose the middle compartment of the bulla, and the round window niche was exposed through a small opening. A small piece of gelfoam soaked with DXM was placed into the round window and bulla (Figure 1B,C). The gelfoam about 1 mm in size soaked with 5 ul DXM (3.75 mg/mL) was administered to the round window and solved within 2 weeks. The incision was closed with nonabsorbable sutures, and the animals were transferred onto a homeothermic blanket at 39.8 °C for the recovery period.

2.3. Auditory Brainstem Response Recording

The animals' auditory function in the surgical ear was assessed by recording the auditory brainstem responses (ABRs) as previously described [22]. In brief, the mice were anesthetized and kept warm with a heating pad in a sound-attenuating chamber. Subdermal needle electrodes were inserted at the vertex (positive), below the pinna of the ear (negative), and at the back (ground) of the mice. Specific stimuli (clicks and 8-, 12-, 16-, 20-, 24-, 28-, and 32-kHz tone bursts) were generated using the SigGen software program (Tucker-Davis Technologies, Gainesville, FL, USA) and delivered to the external auditory canal. The average responses to 1024 stimuli at each frequency were obtained by reducing the sound intensity in 5-dB steps until the threshold was reached. The resulting ABR thresholds were defined as the lowest intensity at which a reproducible deflection in the evoked response trace could be recognized. The ABR wave I peak-to-peak amplitude was computed through an offline analysis of the stored waveforms.

2.4. Population-Base Database

This study employed a retrospective matched-cohort design. We acquired medical records over 40 years old from the National Health Insurance Research Database (NHIRD, an outpatient and hospitalization longitudinal health insurance database in Taiwan) between 1 January 2000 and 31 December 2015, to establish matched cohorts with and without DXM use, using a propensity score method at a ratio of 1:4 by sex, age, and index year. Patients were diagnosed with sensorineural hearing loss (SNHL) according to ICD-9-CM code (389.1x). We excluded all patients with a prior hearing loss diagnosis and those who had been prescribed DXM before 2000.

Gender, age, covariates, and comorbidities were assessed and analyzed in this study. We defined catastrophic illness using the definition of the Ministry of Health and Welfare in Taiwan, which contains 30 categories that the patients can apply for a certificate to become exempt from copayments for healthcare costs related to catastrophic illness [23].

Data used in this study are managed and stored by the Health and Welfare Data Science Center. Researchers can obtain the data through formal application to the Health and Welfare Data Science Center, Department of Statistics, Ministry of Health and Welfare, Taiwan (http://dep.mohw.gov.tw/DOS/np-2497-113.html).

The Institutional Review Board of Tri-Service General Hospital approved this study (TSGHIRB no. 2-105-05-082) and waived the requirement of written informed consent to access the NHIRD.

2.5. Duration of DXM Use and Sensitivity Analysis

The DXM use in this population-based study was prescribed as systemic application. The data of the defined daily dose (DDD) were obtained from the WHO Collaborating Centre for Drug Statistics Methodology (https://www.whocc.no/), and the duration of the usage of DXM was calculated by dividing the cumulative doses by the DDD of DXM. Duration of DXM use was categorized into three subgroups of ≤30, 30~90 and >90 days. The sensitivity test for duration of DXM use was analyzed for risk of developing SNHL using Cox regression model.

2.6. Statistical Analysis

For animal study, the statistical analysis was performed through one-way analysis of variance between control and DXM groups. The results are expressed as the mean ± standard error of the mean. The differences were considered significant at $p < 0.05$.

For human study, all data analyses were performed using IBM SPSS for Windows, version 22.0 (IBM Corp., Armonk, NY, USA). The Chi-square test and Fisher's exact test were used to compare the difference of categorical variables, and Student's t test was used to compare the difference of continuous variables between with DXM use and without DXM use. Multivariate Cox proportional hazards regression was used to determine the risk of SNHL, and the results are presented as a hazard ratio (HR) with 95% confidence interval (CI). The difference in risk of SNHL for patients with or without DXM use was estimated using the Kaplan–Meier method with a log-rank test. A two-tailed p value < 0.05 was considered statistically significant.

3. Results

3.1. Animal Study

A permanent threshold shift was produced after exposure at an SPL of 118 dB for 3 h. A 30 mg/kg dose of DXM was injected intraperitoneally from 2 days before noise exposure to 2 days after noise exposure. We noted that surgery to access the round window did not affect the hearing threshold in mice which was compatible with previous literature [24]. Subsequently, we boosted DXM soaking in the round window the next day after noise exposure. The hearing threshold was measured using the ABRs on day 7, 14, 30, and 60 after noise exposure (Figure 2). From 7 days after noise exposure, the hearing threshold differed significantly between both groups. The hearing threshold was measured as 48.6 ± 2.9 dB through the click ABR in the control group (n = 18) compared with 42.6 ± 7.0 dB in the DXM group (n = 22) on day 60 after noise exposure. The tone burst ABR at frequencies of 24, 28, and 32 K had a significantly lower hearing threshold in the DXM group than in the control group. Although the mice were anesthetized with another NMDA antagonist (ketamine) that may have affected the experimental effect, the difference can be neglected because the same procedure was used in both groups.

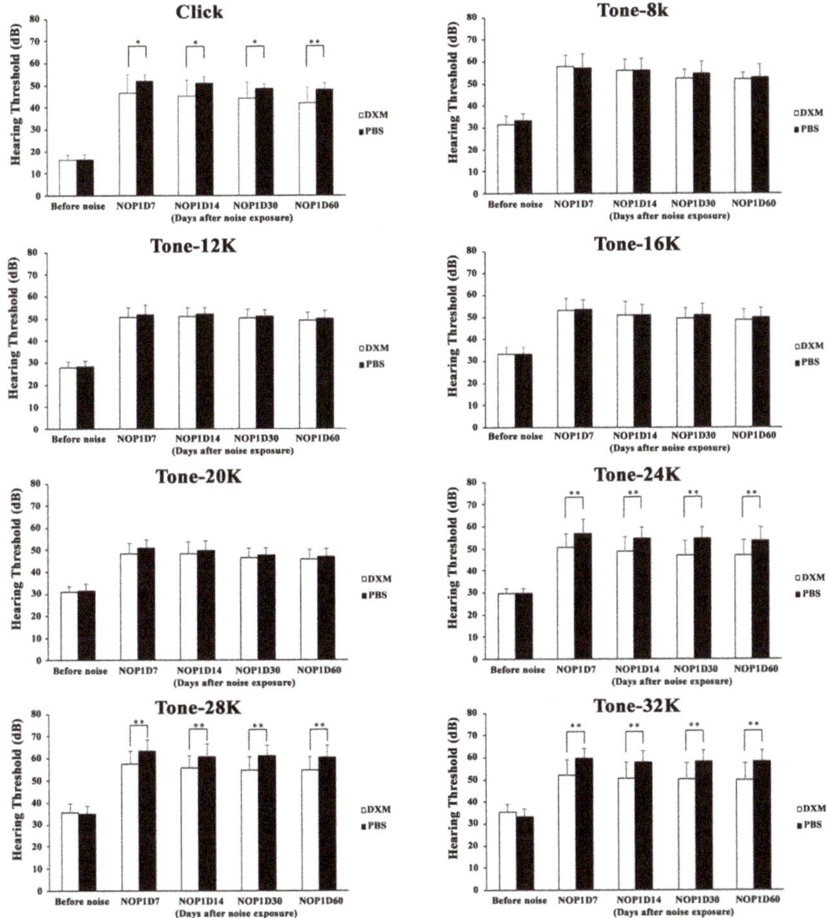

Figure 2. Click and tone burst ABR for hearing evaluation at various time points. The hearing threshold obtained through click ABR (42.6 ± 7.0 dB, n = 22) in the DXM group was significantly (p = 0.002) better than that (48.6 ± 2.9 dB, n = 18) in the control group after noise exposure. In terms of tone burst ABR, the hearing threshold between 24 and 32 kHz in the DXM group was significantly better than that in the control group after noise exposure. * $p < 0.05$, ** $p < 0.01$.

In recent studies, the synapse between inner hair cells and spiral ganglion neurons has been reported to mediate hearing transduction [25]. The amplitude of ABR wave I can be represented as the number of synaptic ribbons [6]. We also analyzed the raw ABR data from the mice on day 60 for wave I amplitude. The data obtained through the click ABR indicated that the morphology of waveforms were better and wave I amplitude was higher in the DXM group compared with the PBS group (Figure 3). This result may reflect the potential effect of DXM treatment with preserved synaptic complexes.

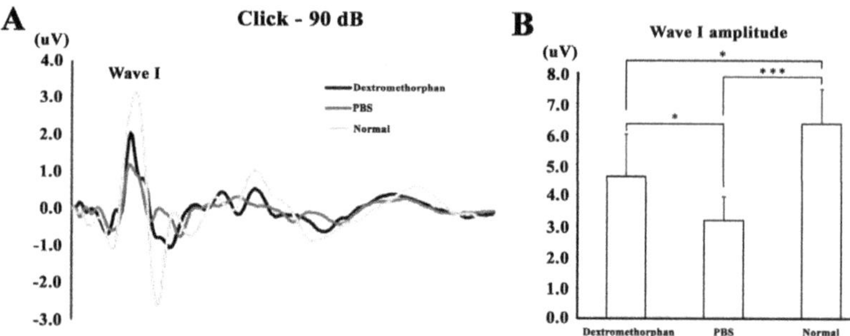

Figure 3. (**A**) ABR waveforms appeared in normal mice without noise exposure, and in the DXM and PBS mice groups with noise exposure. (**B**) ABR wave I amplitude was higher in the DXM group compared with the PBS group. * $p < 0.05$, *** $p < 0.001$.

3.2. Population-Based Human Study

Based on the data used in this study from 1 January 2000 to 31 December 2015, 39,895 individuals with DXM use were included and a matched 159,580 individuals without DXM use were selected as the control group (Figure 4). At the end of the follow-up period (Table 1), 175 individuals with DXM use (0.44%, 175/39,895) and 1675 without DXM use (1.05%, 1675/159,580) had developed SNHL, indicating a significantly lower incidence of hearing loss among those with DXM use ($p < 0.001$). The average follow-up period was 9.90 ± 9.28 years, and the average period for developing hearing loss was 4.87 ± 5.54 years. Significantly lower percentages of ischemia heart disease (IHD) (12.94% vs. 13.70%; $p = 0.021$) and depression (0.48% vs. 0.65%; $p = 0.026$) were found in individuals with DXM use, compared with the control group.

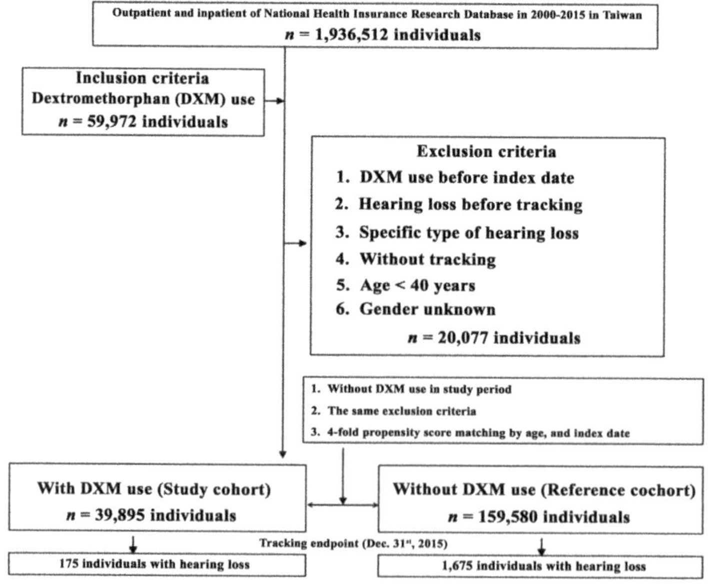

Figure 4. Flowchart of study sample selection.

Table 1. Characteristics of study in the endpoint during the follow-up period from 1 January 2000 to 31 December 2015.

DXM Use	With		Without		p
	n	%	n	%	
Total	39,895	20.00	159,580	80.00	
Hearing loss	175	0.44	1675	1.05	<0.001
Variables					
Gender					0.999
Male	20,121	50.43	80,484	50.43	
Female	19,774	49.57	79,096	49.57	
Age (years)	55.28 ± 26.45		54.70 ± 25.51		<0.001
Catastrophic illness	2671	6.70	12,104	7.58	<0.001
DM	3884	9.74	16,011	10.03	0.076
HTN	9112	22.84	35,142	22.02	<0.001
Depression	1097	2.75	4048	2.54	0.016
Insomnia	1845	4.62	6822	4.27	0.002
Stroke	2111	5.29	8705	5.45	0.197
CKD	4344	10.89	17,035	10.67	0.217
Hyperlipidaemia	1184	2.97	4499	2.82	0.111
Epilepsy	429	1.08	1375	0.86	<0.001
AID	3245	8.13	13,401	8.40	0.088
IHD	2675	6.71	12,604	7.90	<0.001
COPD	2973	7.45	12,801	8.02	<0.001
Pneumonia	5671	14.21	20,754	13.01	<0.001
Head injury	6881	17.25	26,279	16.47	<0.001
Asthma	4213	10.56	17,987	11.27	<0.001
Alcohol abuse/dependence	381	0.96	1275	0.80	0.002
Tobacco abuse/dependence	291	0.73	1248	0.78	0.282
CLD	4041	10.13	16,124	10.10	<0.001
Parkinson's disease	900	2.26	3251	2.04	0.006
Urbanization level					<0.001
1 (The highest)	12,121	30.38	44,026	27.10	
2	13,986	35.06	54,521	33.57	
3	5111	12.81	20,782	12.79	
4 (The lowest)	8677	21.75	43,101	26.54	
Level of care					<0.001
Hospital center	14,562	36.50	46,128	28.91	
Regional hospital	15,756	39.49	65,117	40.81	
Local hospital	9577	24.01	48,335	30.29	

p: Chi-square/Fisher exact test on category variables and t-test on continue variables; DM = diabetes mellitus; HTN = hypertension; CKD = chronic kidney disease; AID = autoimmune diseases; IHD = ischemia heart disease; COPD = chronic obstruction pulmonary disease; CLD = chronic liver disease.

The cumulative incidence curve of SNHL for the cohort with DXM use was significantly lower than for the control cohort, following adjustment for age and other variables (Figure 5; log-rank test, $p < 0.001$).

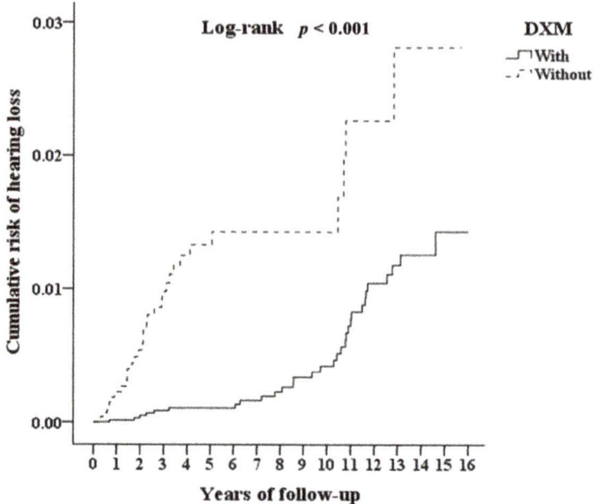

Figure 5. Kaplan–Meier for cumulative risk of hearing loss among individuals aged > 40 with or without DXM use.

After adjustment for age, sex, and comorbidities in the Cox proportional hazard regression, DXM use was still significantly associated with a decreased risk of hearing loss, with an adjusted HR of 0.725 (95% CI, 0.624–0.803; $p < 0.001$; Supplementary Table S1). Patients with catastrophic illness, older patients, diabetes mellitus, hypertension, stroke, chronic kidney disease, autoimmune diseases, IHD, pneumonia, head injury, and chronic liver disease had a significantly higher risk of SNHL.

We investigated the sensitivity test for duration of DXM use and the test demonstrated a lower risk of developing SNHL was associated with longer duration of DXM use (Table 2).

Table 2. Factors of hearing loss among different duration of DXM by using Cox regression.

DXM Dose	Population	Events	PYs	Rate (per 10^5 PYs)	Adjusted HR	95% CI	95% CI	p
Without	159,580	1675	2,012,465.11	83.23	Reference			
With	39,895	175	513,401.02	34.09	0.725	0.624	0.803	<0.001
1–30 days	8423	43	114,267.35	37.63	0.797	0.684	0.888	<0.001
31–90 days	19,450	88	248,640.12	35.39	0.756	0.642	0.834	<0.001
≥91 days	12,022	44	150,493.55	29.24	0.622	0.531	0.697	<0.001

PYs = Person-years; Adjusted HR = Adjusted Hazard ratio: Adjusted for the variables listed in Table 1; CI = confidence interval; DXM = dextromethorphan.

4. Discussion

Our data demonstrated and explored that DXM, an NMDA antagonist, had the ability to attenuate noise-induced hearing loss (NIHL) by improving the hearing threshold and wave I amplitude, as indicated through ABR testing, and also potentially decrease the risk of SNHL after adjusting by Cox regression analysis in a 16-year follow-up nationwide population-based study. To the best of our knowledge, this is the first report to explore the preventive effect of DXM use in hearing loss in both animal and human study.

Some NMDA antagonists have been used to investigate the protective effect in the inner ear using animal models of noise [11,12,14], ischemia [17,26,27], and ototoxicity [10,12,28]. Jäger et al. demonstrated that dizocilpine maleate (MK-801) had a protective effect against NIHL, but the effect was limited to a specific frequency [11]. Duan et al. also presented MK-801 as a means to substantially limit both NIHL and swelling of dendrites under inner hair cells [12]. Bing et al. reported that esketamine hydrochloride gel (AM-101) may reduce the noise-induced loss of synaptic ribbons, but it had no protective effect against NIHL [14]. Tabuchi et al. reported that ketamine and dextromethorphan, but not MK-801 have protective effect on cochlear ischemia dysfunction [17]. MK-801 and ifenprodil resulted in the protection of aminoglycoside-induced ototoxicity [12,28]. There was still sparse study to explore the effect of DXM use in common hearing loss animal models.

Given that NMDA receptor antagonists tend to have negative central nervous system side effects, such as hallucinations or anesthesia, their clinical potential has been limited [14]. In the present study, we demonstrated that DXM, widely used and available in numerous over-the-counter cough and cold preparations worldwide [16], also has a protective effect against NIHL. The result was compatible to our previous report in a NIHL rat model using 4-[^{18}F]-ADAM/micro-PET and showed DXM prevent hearing loss and preserved brain serotonin transporter function [21]. This effect also has been proven previously that DXM moderately ameliorated the compound action potential threshold shift in cochlear dysfunction induced by transient ischemia [17].

Based on aforementioned results, DXM has the potential to be a promising therapeutic intervention, we directly explored the association between DXM use and the risk of SNHL using Taiwan's NHIRD database. After adjusting variables by multiple Cox regression analysis, the use of DXM significantly reduced the risk of SNHL compared with matched-cohort control group. DXM use is frequently prescribed for patients when they might catch cold in Taiwan. We used cumulative DDD (cDDD) of DXM to analyze the association between DXM use and risk of SNHL. The result revealed longer duration of DXM use (larger amount of cDDD) had more significantly decreased the risk of SNHL. Herein, we further provided the evidence to support the finding that the use of DXM has a strong impact in preventing SNHL.

The mechanism for the preventive effect of DXM from hearing loss may be through the inhibition of glutamate-induced excitotoxicity at the highly active synapse due to NMDA receptor activation, dendritic swelling, and the production of reactive oxygen species [9,29]. Glutamate-induced excitotoxicity was recently thought as an instigating factor of cochlear synaptopathy [8,9]. The use of DXM can have the capacity to protect synaptic damage in acute noise-exposed hearing loss.

In recent years, a hidden hearing loss theory has advocated that synaptic loss is the primary pathology even with only temporary threshold shifts, and that this synaptic loss is independent from both IHC and SGN loss [6,30]. Cochlear synaptopathy may contribute to hearing impairment in millions of people [8,31]. If these damaged synaptic connections can be maintained or restored by any potential drugs or other therapeutic management, it could undoubtedly improve the hearing function [32]. Some reports have evidenced that neurotrophins, such as neurotrophin-3 and brain-derived neurotrophic factors, can offer protective effects against noise trauma in animals [24,33,34]. DXM would be an effective and alternative drug to apply in further human studies on NIHL (e.g., recreational or military gun shooting) or age-related SNHL.

Our study had some limitations. In animals, it was not clear which approach (intraperitoneal injection or round window soaking) resulted in the therapeutic effect because we combined both approaches and we did not assess the immunohistochemistry staining for synaptic ribbon. In humans, we were unable to reduce non DXM contributing factors in the retrospective-matched cohort study. This is because: (1) we cannot assess a patient's history and cause of SNHL according to the diagnosis by using the ICD-9-CM code (389.1x), (2) the severity of hearing threshold of individuals as determined by audiometry was not available in the current NHIRD database, and (3) the timing of DXM application was inaccessible as it was not prescribed for treatment of SNHL. All of these factors may have contributed to the otoprotective effect of DXM in this study. Future clinical trials would have to sufficiently reduce non DXM contributing factors to confirm our findings.

5. Conclusions

Our study demonstrated that DXM significantly attenuated noise-induced hearing loss and may have a protective effect against SNHL. The effect may be through synaptic regulation. Therefore, to develop a novel effect of pharmaceutical medicine to prevent cochlear synaptopathy is practical to the hearing preservation program. DXM can be an alternative medication applying in future clinical trials for SNHL.

Supplementary Materials: The following are available online at http://www.mdpi.com/1660-4601/17/17/6336/s1, Table S1: Factors of hearing loss by using Cox regression.

Author Contributions: Conceptualization, H.-C.C. and C.-H.W.; methodology, H.-C.C., C.-H.C. and W.-C.C.; software, C.-H.C. and W.-C.C.; validation, H.-C.C., C.-H.W. and W.-C.C.; formal analysis, C.-H. C. and W.-C.C.; resources, H.-C.C. and W.-C.C.; data curation, H.-C.C., C.-H.W. and W.-C.C.; writing—original draft preparation, H.-C.C.; writing—review and editing, C.-P.S., I.-H.L., Y.-C.L., Y.-Y.L. and C.-Y.K.; supervision, C.-H.W.; project administration, H.-C.C. and W.-C.C.; funding acquisition, H.-C.C. All authors have read and agreed to the published version of the manuscript.

Funding: This work, in part, received grants from the Ministry of Science and Technology, Taiwan (MOST 108-2314-B-016-037 to H.C. Chen), and Tri-Service General Hospital (TSGH-D-109053 to H.C. Chen). The funders had no role in the study design, data collection and analysis, decision to publish, or preparation of the manuscript.

Conflicts of Interest: The authors declare no conflict of interest.

References

1. Wilson, B.S.; Tucci, D.L.; Merson, M.H.; O'Donoghue, G.M. Global hearing health care: New findings and perspectives. *Lancet* **2017**, *390*, 2503–2515. [CrossRef]
2. Disease, G.B.D.; Injury, I.; Prevalence, C. Global, regional, and national incidence, prevalence, and years lived with disability for 328 diseases and injuries for 195 countries, 1990–2016: A systematic analysis for the Global Burden of Disease Study 2016. *Lancet* **2017**, *390*, 1211–1259. [CrossRef]
3. Wang, J.; Puel, J.L. Toward Cochlear Therapies. *Physiol. Rev.* **2018**, *98*, 2477–2522. [CrossRef] [PubMed]
4. Gazia, F.; Abita, P.; Alberti, G.; Loteta, S.; Longo, P.; Caminiti, F.; Gargano, R. NICU Infants & SNHL: Experience of a western Sicily tertiary care centre. *Acta Med. Mediterr.* **2019**, *35*, 1001–1007.
5. Valero, M.D.; Burton, J.A.; Hauser, S.N.; Hackett, T.A.; Ramachandran, R.; Liberman, M.C. Noise-induced cochlear synaptopathy in rhesus monkeys (*Macaca mulatta*). *Hear. Res.* **2017**, *353*, 213–223. [CrossRef]
6. Kujawa, S.G.; Liberman, M.C. Adding insult to injury: Cochlear nerve degeneration after "temporary" noise-induced hearing loss. *J. Neurosci.* **2009**, *29*, 14077–14085. [CrossRef]
7. Hickox, A.E.; Larsen, E.; Heinz, M.G.; Shinobu, L.; Whitton, J.P. Translational issues in cochlear synaptopathy. *Hear. Res.* **2017**, *349*, 164–171. [CrossRef]
8. Moser, T.; Predoehl, F.; Starr, A. Review of hair cell synapse defects in sensorineural hearing impairment. *Otol. Neurotol.* **2013**, *34*, 995–1004. [CrossRef]
9. Liberman, M.C.; Kujawa, S.G. Cochlear synaptopathy in acquired sensorineural hearing loss: Manifestations and mechanisms. *Hear. Res.* **2017**, *349*, 138–147. [CrossRef]
10. Hong, J.; Chen, Y.; Zhang, Y.; Li, J.; Ren, L.; Yang, L.; Shi, L.; Li, A.; Zhang, T.; Li, H.; et al. *N*-Methyl-D-Aspartate Receptors Involvement in the Gentamicin-Induced Hearing Loss and Pathological Changes of Ribbon Synapse in the Mouse Cochlear Inner Hair Cells. *Neural Plast.* **2018**, *2018*, 3989201. [CrossRef]
11. Jager, W.; Goiny, M.; Herrera-Marschitz, M.; Brundin, L.; Fransson, A.; Canlon, B. Noise-induced aspartate and glutamate efflux in the guinea pig cochlea and hearing loss. *Exp. Brain Res.* **2000**, *134*, 426–434. [CrossRef] [PubMed]
12. Duan, M.; Agerman, K.; Ernfors, P.; Canlon, B. Complementary roles of neurotrophin 3 and a *N*-methyl-D-aspartate antagonist in the protection of noise and aminoglycoside-induced ototoxicity. *Proc. Natl. Acad. Sci. USA* **2000**, *97*, 7597–7602. [CrossRef] [PubMed]
13. Duan, M. Glutamate receptor antagonist and neurotrophin can protect inner ear against damage. *J. Otol.* **2009**, *4*, 26–33.
14. Bing, D.; Lee, S.C.; Campanelli, D.; Xiong, H.; Matsumoto, M.; Panford-Walsh, R.; Wolpert, S.; Praetorius, M.; Zimmermann, U.; Chu, H.; et al. Cochlear NMDA receptors as a therapeutic target of noise-induced tinnitus. *Cell. Physiol. Biochem. Int. J. Exp. Cell. Physiol. Biochem. Pharmacol.* **2015**, *35*, 1905–1923. [CrossRef]
15. Furman, A.C.; Kujawa, S.G.; Liberman, M.C. Noise-induced cochlear neuropathy is selective for fibers with low spontaneous rates. *J. Neurophysiol.* **2013**, *110*, 577–586. [CrossRef]

16. Nguyen, L.; Thomas, K.L.; Lucke-Wold, B.P.; Cavendish, J.Z.; Crowe, M.S.; Matsumoto, R.R. Dextromethorphan: An update on its utility for neurological and neuropsychiatric disorders. *Pharmacol. Ther.* **2016**, *159*, 1–22. [CrossRef] [PubMed]
17. Tabuchi, K.; Tsuji, S.; Wada, T.; Ito, Z.; Hara, A.; Kusakari, J. Effect of ketamine, dextromethorphan, and MK-801 on cochlear dysfunction induced by transient ischemia. *Ann. Otol. Rhinol. Laryngol.* **2002**, *111*, 44–49. [CrossRef]
18. Werling, L.L.; Lauterbach, E.C.; Calef, U. Dextromethorphan as a potential neuroprotective agent with unique mechanisms of action. *Neurologist* **2007**, *13*, 272–293. [CrossRef]
19. Rizzo, S.; Bentivegna, D.; Thomas, E.; La Mattina, E.; Mucia, M.; Salvago, P.; Sireci, F.; Martines, F. Sudden Sensorineural Hearing Loss, an Invisible Male: State of the Art. In *Hearing Loss: Etiology, Management and Societal Implications*; Hughes, J.D., Ed.; Nova Science Publishers Inc.: New York, NY, USA, 2017; pp. 75–86.
20. Chen, J.; Shi, R. Current advances in neurotrauma research: Diagnosis, neuroprotection, and neurorepair. *Neural Regen. Res.* **2014**, *9*, 1093–1095. [CrossRef]
21. Liu, C.T.; Huang, Y.S.; Chen, H.C.; Ma, K.H.; Wang, C.H.; Chiu, C.H.; Shih, J.H.; Kang, H.H.; Shiue, C.Y.; Li, I.H. Evaluation of brain SERT with 4-[(18)F]-ADAM/micro-PET and hearing protective effects of dextromethorphan in hearing loss rat model. *Toxicol. Appl. Pharmacol.* **2019**, *378*, 114604. [CrossRef]
22. Shih, C.P.; Chen, H.C.; Lin, Y.C.; Chen, H.K.; Wang, H.; Kuo, C.Y.; Lin, Y.Y.; Wang, C.H. Middle-ear dexamethasone delivery via ultrasound microbubbles attenuates noise-induced hearing loss. *Laryngoscope* **2019**, *129*, 1907–1914. [CrossRef] [PubMed]
23. Yang, N.P.; Lee, Y.H.; Chung, C.Y.; Hsu, J.C.; Yu, I.L.; Chang, N.T.; Chan, C.L. Comparisons of medical utilizations and categorical diagnoses of emergency visits between the elderly with catastrophic illness certificates and those without. *BMC Health Serv. Res.* **2013**, *13*, 152. [CrossRef]
24. Suzuki, J.; Corfas, G.; Liberman, M.C. Round-window delivery of neurotrophin 3 regenerates cochlear synapses after acoustic overexposure. *Sci. Rep.* **2016**, *6*, 24907. [CrossRef] [PubMed]
25. Liberman, M.C. Noise-induced and age-related hearing loss: New perspectives and potential therapies. *F1000Research* **2017**, *6*, 927. [CrossRef]
26. Pujol, R.; Puel, J.L.; Eybalin, M. Implication of non-NMDA and NMDA receptors in cochlear ischemia. *Neuroreport* **1992**, *3*, 299–302. [CrossRef]
27. Lipton, S.A.; Rosenberg, P.A. Excitatory amino acids as a final common pathway for neurologic disorders. *N. Engl. J. Med.* **1994**, *330*, 613–622. [CrossRef]
28. Basile, A.S.; Huang, J.M.; Xie, C.; Webster, D.; Berlin, C.; Skolnick, P. *N*-methyl-D-aspartate antagonists limit aminoglycoside antibiotic-induced hearing loss. *Nat. Med.* **1996**, *2*, 1338–1343. [CrossRef]
29. Ruel, J.; Chabbert, C.; Nouvian, R.; Bendris, R.; Eybalin, M.; Leger, C.L.; Bourien, J.; Mersel, M.; Puel, J.L. Salicylate enables cochlear arachidonic-acid-sensitive NMDA receptor responses. *J. Neurosci.* **2008**, *28*, 7313–7323. [CrossRef]
30. Kobel, M.; Le Prell, C.G.; Liu, J.; Hawks, J.W.; Bao, J. Noise-induced cochlear synaptopathy: Past findings and future studies. *Hear. Res.* **2017**, *349*, 148–154. [CrossRef]
31. Mepani, A.M.; Kirk, S.A.; Hancock, K.E.; Bennett, K.; de Gruttola, V.; Liberman, M.C.; Maison, S.F. Middle Ear Muscle Reflex and Word Recognition in "Normal-Hearing" Adults: Evidence for Cochlear Synaptopathy? *Ear Hear.* **2020**, *41*, 25–38. [CrossRef]
32. Cunningham, L.L.; Tucci, D.L. Restoring synaptic connections in the inner ear after noise damage. *N. Engl. J. Med.* **2015**, *372*, 181–182. [CrossRef] [PubMed]
33. Wan, G.; Gomez-Casati, M.E.; Gigliello, A.R.; Liberman, M.C.; Corfas, G. Neurotrophin-3 regulates ribbon synapse density in the cochlea and induces synapse regeneration after acoustic trauma. *eLife* **2014**, *3*. [CrossRef] [PubMed]
34. Sly, D.J.; Campbell, L.; Uschakov, A.; Saief, S.T.; Lam, M.; O'Leary, S.J. Applying Neurotrophins to the Round Window Rescues Auditory Function and Reduces Inner Hair Cell Synaptopathy After Noise-induced Hearing Loss. *Otol. Neurotol.* **2016**, *37*, 1223–1230. [CrossRef] [PubMed]

International Journal of
Environmental Research and Public Health

Article

Voice-Related Quality of Life in Post-Laryngectomy Rehabilitation: Tracheoesophageal Fistula's Wellness

Salvatore Cocuzza [1], Antonino Maniaci [1,*], Calogero Grillo [1], Salvatore Ferlito [1], Giacomo Spinato [2], Salvatore Coco [1], Federico Merlino [1], Giovanna Stilo [1], Giovanni Paolo Santoro [3], Giannicola Iannella [4,5], Claudio Vicini [4,6] and Ignazio La Mantia [1]

1 Department of Medical and Surgical Sciences and Advanced Technologies "GF Ingrassia", ENT Section, University of Catania, 95100 Catania, Italy; s.cocuzza@unict.it (S.C.); grillo@unict.it (C.G.); ferlito@unict.it (S.F.); salvo.coco93@gmail.com (S.C.); federico.merlino93@gmail.com (F.M.); giovastilo@libero.it (G.S.); igolama@gmail.com (I.L.M.)
2 Section of Otorhinolaryngology, University of Padova, 31100 Treviso, Italy; giacomo.spinato@unipd.it
3 Head and Neck and Robotic Surgery, Department of Experimental and Clinical Medicine, University of Florence, 50134 Florence, Italy; gp.santoroo@gmail.com
4 Head and Neck Department, ENT & Oral Surgery Unit, G.B. Morgagni, L. Pierantoni Hospital, 47121 Forlì, Italy; giannicola.iannella@uniroma1.it (G.I.); claudio@claudiovicini.com (C.V.)
5 Department of Organi di Senso, University "Sapienza", 00185 Rome, Italy
6 Department of Otolaryngology, Head and Neck Surgery, University of Ferrara, 44121 Ferrara, Italy
* Correspondence: tnmaniaci29@gmail.com

Received: 6 May 2020; Accepted: 24 June 2020; Published: 26 June 2020

Abstract: (1) Introduction: Laryngeal cancer is one of the most common types of cancer affecting the upper aerodigestive tract. Despite ensuring good oncological outcome in many locoregionally advanced cases, total laryngectomy is associated with relevant physical and psychological sequelae. Treatment through tracheo-esophageal speech, if promising, can lead to very variable outcomes. Not all laryngectomee patients with vocal prosthesis benefit from the same level of rehabilitation mainly due to the development of prosthetic or fistula related problems. The relating sequelae in some cases are even more decisive in the patient quality of life, having a higher impact than communicational or verbal skills. (2) Material and Methods: A retrospective study was conducted on 63 patients initially enrolled with a history of total laryngectomy and voice rehabilitation, treated at the University Hospital of Catania from 1 January 2010 to 31 December 2018. Quality of life (QoL) evaluation through validated self-administrated questionnaires was performed. (3) Results: The Voice-Related Quality of Life questionnaire revealed significantly better outcomes in both socio-emotional and functional domains of the tracheoesophageal patient group compared to the esophageal group ($p = 0.01$; $p = 0.01$, respectively), whereas in the Voice Handicap Index assessment, statistically significant scores were not achieved ($p = 0.33$). (4) Discussion: The significant differences reported through the V-RQOL and Voice Handicap Index scales in the presence of fistula related problems and device lifetime reduction when compared to the oesophageal speech group have demonstrated, as supported by the literature, a crucial role in the rehabilitative prognosis. (5) Conclusions: The criteria of low resistance to airflow, optimal tracheoesophageal retention, prolonged device life, simple patient maintenance, and comfortable outpatient surgery are the reference standard for obtaining good QoL results, especially over time. Furthermore, the correct phenotyping of the patient based on the main outcomes achieved at clinical follow-up guarantees the primary objective of the identification of a better quality of life.

Keywords: quality of life assessment; tracheoesophageal speech; tracheo-esophageal puncture

1. Introduction

Laryngeal cancer is one of the most common types of cancer affecting the upper aerodigestive tract, and accounts for about 177,000 new cases per year worldwide with an estimated 94,800 cancer deaths annually [1,2]. These patients who undergo head and neck surgery often experience challenges in communication, which may impact the quality of life (QoL) [2–4]. The World Health Organization (WHO) defines health as *"A state of complete physical, mental, and social well-being not merely the absence of disease"* and the QoL as an *"individual's perception of their position in life in the context of the culture and value systems in which they live and concerning their goals, expectations, standards and concerns"* [5].

Despite ensuring good oncological outcome in many locoregionally advanced cases, total laryngectomy (TL) is associated with relevant physical and psychological sequelae, affecting essential life functions such as breathing, swallowing, and oral communication. Permanent tracheostomy and loss of natural voice worsen patients' QoL, resulting in social stigma and consequent psychological discomfort.

To date, among the various voice recovery solutions, the tracheo-esophageal puncture (TEP) with the insertion of voice prosthesis (VP) is the method generally recognized as a gold standard procedure for speech restoration after laryngectomy.

Technological advancements have been such that today, indwelling prostheses are designed to meet the criteria of low airflow resistance and optimal retention in the tracheoesophageal party wall, have prolonged device lifetime, simple maintenance by the patient, and comfortable outpatient replacement, with success rates classically from 40 to 90% with excellent voice quality [6–9]. More recent findings by Souza et al. in a study population of 95 patients reported the overall quality of life in voice prosthesis from good to excellent in 83.2% of cases, with better results compared to esophageal voice [8].

Several studies have previously performed an assessment of QoL and degree of satisfaction in TL patients after VP rehabilitation, demonstrating its good efficacy in highly motivated subjects, even over time [9,10].

Despite this, it is necessary to consider, however, that not all laryngectomee patients with vocal prosthesis benefit from the same level of rehabilitation. Indeed, the spread of known prosthetic device lifetime or fistula related problems, which is a more or less decisive way, has an impact on the level and the quality of speech rehabilitation. A drastically reduced device lifetime, periprosthetic leakage, recurrent tracheoesophageal granulomas, and poor vocal performance forces the patient to undergo a higher number of medical-surgical procedures, often aggressive, with potential secondary psychophysical discomfort. The following rehabilitation level often results in not being optimal and autonomous, leading to a possible failure and closure of the phonatory fistula. To the above, numerous reports have identified pathological or uncontrolled gastroesophageal reflux, the chronic mucosal outcomes of radiotherapy, as known etiopathogenetic factors that, alone or in association, have a decisive impact on voice prosthesis outcomes [11].

Therefore, the presence of good communication indices is not necessarily related to the degree of satisfaction of the prosthetic treatment and the related socio-emotional consequences [12].

To this end, through the use of dedicated evaluation questionnaires within a court of laryngectomies and prosthetic patients, we retrospectively evaluated the variability of the related QoL based on specific indicators. Categorization, according to the presence or absence of prosthetic or fistula-related disorders comparing the obtained results with an esophageal speech group was performed.

2. Material and Methods

A retrospective study on the QoL of laryngectomized patients with a different degree of tracheoesophageal voice prosthesis rehabilitation was performed. The results were then compared with a group of patients with esophageal voice rehabilitation.

2.1. Selection Process

A total of 63 patients initially treated at the University Hospital of Catania from 1 January 2010 to 31 December 2018 were initially considered.

Patients were enrolled in compliance with the following inclusion/exclusion criteria:

1. Inclusion Criteria: Tracheoesophageal speech utilizing a voice prosthesis after laryngectomy procedures for laryngeal tumor (both primary and secondary TEP performed); esophageal speech; clinical-instrumental follow-up after TEP performed ≥ 10 years.
2. Exclusion Criteria: patients with evident local recurrence of pathology; the presence of comorbidities with significant impact on the patient's QoL (neurological, cerebrovascular, or cardiovascular accident, myopathy); not related psychological diseases occurred after surgery; exitus for external causa before completing the study.

2.2. Patients Population

A total of nine patients were excluded due to the selection criteria while 54 patients were successfully interviewed, and enrolled. The enrolled patients consisted of 47 males and seven females, aged from 53 to 78 years (mean age: 64.7 years). TEP was performed as a primary procedure in 10 cases while as a secondary one in 29 cases, generally almost six months after TL (6.1 ± 1.4 months). Instead, esophageal speech rehabilitation was achieved in 15 patients.

All sociodemographic and clinical characteristics such as the mean age, patient sex, mean follow up, TEP procedures, and the history of radiotherapy treatment are summarized in Table 1.

Table 1. Demographic and clinical features.

Characteristic	No. (%)/Range	Mean	SD
Sex			
Male	47 (87%)		
Female	7 (13%)		
Age	53–78 y	64.7 y	±7.58 y
Mean follow up, y		11.2 y	±1.65 y
T Stage			
III	35 (64.8%)		
IV	19 (35.2%)		
Neck dissection			
Yes	39 (72.2%)		
No	15 (27.8%)		
Radiation			
Irradiated	35 (64.8%)		
Not irradiated	19 (35.2%)		
TEP Procedure			
Primary	10 (18.5%)		
Secondary	29 (53.7%)		
Esophageal Voice	15 (27.8%)		

Abbreviations: T stage = tumor stage; TEP = tracheoesophageal prosthesis.

All data were collected from the same two surgeons (S.C. and I.L.M.) who followed the post-operative control up to its fulfilment.

According to prosthesis outcomes and fistula related disorders observed at clinical follow-up, we classified patients into two patient groups:

1. Group 1 including all patients treated with tracheoesophageal voice rehabilitation.
2. Group 2 defined by patients performing esophageal voice rehabilitation.

Moreover, in the second part of the analysis, based on the main complications recorded in the TEP group, we identified two phenotypic subgroups and assessed specific QoL results for each one:

- Prosthetic disorders group (PD), defined by a prosthetic device lifetime ≤3 months; and
- Fistula-related disorders group (FRD), defined by the presence of subsequent complications such as periprosthetic leakage, macro fistula, recurrent tracheoesophageal granuloma).

2.3. Outcome Assessment

Patients who respected the selection criteria underwent subjective and objective evaluations:

1. Quality of life Assessment obtained through the administration of the Voice-Related Quality of Life questionnaire (VR-QoL);
2. Vocal performance evaluation through the Voice Handicap Index (VHI);
3. Percentage of annual complications (failure due to periprosthetic leakage, presence of tracheoesophageal granulation tissue, spontaneous dislodged prosthesis, macro fistula);
4. Median device lifetime duration per year, subsequently divided in each sub-category, through consecutive follow up was recorded.

2.4. Quality of Life (QoL) Assessment

This was carried out through the use of a specific questionnaire administered in each of the two main identified groups, the TEP group (total, prosthesis and fistula related disorders) and ES group.

The V-RQoL questionnaire is a self-administered assessment tool that evaluates the subjective burden elicited by a voice disorder [13]. It is made up 10 statements on aspects related to voice through the emotion, physical, and functional domains with a score of 0–50. Therefore, the degree of severity of the score is directly proportional to the total numerical sum, detecting the poor quality of life for high total values obtained.

The score has two domains, of which six items evaluate the physical functioning (PF) and four the social-emotional (SE).

2.5. Subjective Voice Disorders Assessment

The VHI is made up of three parts concerning the emotional, physical, and functional component, respectively, each composed of 10 items. Each item is marked on a 4-point scale, and the score ranges from 0 to 120, where 120 represents the maximum perceived disability. The score resulting is divided into mild impact (0–40 points), moderate impact (41–60), and severe impact (score > 60).

2.6. Statistical Analysis and Ethical Statement

Data analysis was performed using IBM SPSS Statistics for Windows, IBM Corp. Released 2017, Version 25.0. Armonk, NY: IBM Corp. Descriptive statistics were reported on average ± standard deviation or proportion. Data normality was assessed using the Kolmogorov–Smirnov test of normality. The T-test for paired samples was used to determine the difference between observations. The Mann–Whitney U test was performed to analyze group differences. The tests were two-tailed, and a p-value of < 0.05 was considered as statistically significant.

The study protocol was approved by the ethics committee of the involved Institution (CE Catania 2; Prot. N. 298/BE). Participants were informed and gave written informed consent of the purpose and procedures of the study, which was conducted according to the Declaration of Helsinki.

3. Results

A total number of 54 patients were successfully interviewed after initial selection (47 males and seven females). After the selection process, 39/54 (72.2%) who performed TEP rehabilitation were included, whereas 15/54 (27.8%) among the subjects rehabilitated with ES voice were selected. The ATOS medical puncture set and prosthesis system were used in all patients.

The timing of the TEP procedure performed was primary in 10 (18.5%) patients while secondary in 29 (53.7%) patients.

There were no significant differences between the two groups regarding the demographic and clinical data recorded as summarized in Table 1.

3.1. Voice Prosthesis Sequelae and Device Lifetime

At the follow-up conducted in the TEP group, a leakage was detected in 18/39 (46.1%) cases who underwent fistulization. In particular, eight (20.5%) (Group PD) patients presented leakage through the prosthesis, associated with an average device lifetime fewer than 90 days, while 10/39 (25.6%) (Group FRD) patients presented peri prosthesis leakage. In this last group, the related sequelae identified were granulation tissue in 7/39 (17.9%) whereas there were changes in fistula size in 3/39 (7.7%) (Table 2).

Table 2. Percentage of complications per patient.

Complications TEP Group	No. (%) y
Prosthesis leak	
Through Peri	8/39 (20.5%) 10/39 (25.6%)
Granulation's tissue	7/39 (17.9%)
Fistula size changes	3/39 (7.7%)
Device lifetime	Mean days (SD)
H group FT group PD group	97.4 ± 8.8 days 91.3 ± 6.5 days 61.9 ± 9.6 days

Abbreviations: H = healthy; FT = Fistula type; PD = Prosthesis disorder group.

The mean device lifetime recorded in TEP patients was 83.53 ± 8.3. Through the TEP group sub-typing based on the detected complications, we found a 3-month lower device lifetime only in the patients with prosthetic disorders (PD) compared to the healthy TEP patients (H) and fistula related (FR) ones, respectively (PD 61.9 ± 9.6 days; H 97.4 ± 8.8 days; FR 91.3 ± 6.5 days) (Table 2).

3.2. Voice-Related Quality of Life Questionnaire Assessment

The VrQoL questionnaire revealed significantly better outcomes in both *Socio-Emotional* and *Functional* domains of the TEP patient group compared to the EV group ($p = 0.01$; $p = 0.01$, respectively) (Table 3).

Table 3. Comparison of V-RQoL outcomes.

VrQoL	Tracheoesophageal Voice Prosthesis (TEP)	Voice Prosthesis Disorders (PD)	Fistula Related Disorders (FRD)	Esophageal Speech (EV)
No. Patients	39	8	10	15
Socio-Emotional	4.15 [a] ± 2.23	3.47 ± 0.54	7.18 ± 2.22	4.78 [a] ± 1.03
Functional	4.57 [b] ± 2.48	4.16 ± 1.19	8.33 ± 1.23	5.98 [b] ± 1.18
Total	8.73 [c] ± 4.71	7.63 ± 1.73 [d]	15.51 ± 3.45 [e]	10.76 [c, d, e] ± 2.21

Abbreviations: VrQoL = Voice Related Quality of Life; TEP = tracheoesophageal prosthesis; No. patients = number of patients. Comparison [a] Socio-Emotional TEP vs. EV $p = 0.01$; [b] Functional TEP vs. EV $p = 0.01$; [c] Total TEP vs. Total EV $p = 0.01$; [d] Total PD vs. EV $p = 0.002$; [e] Total FRD vs. EV $p = 0.0007$.

The evaluation of the subgroups divided by specific complications showed better results in the group of prosthetic disorders, despite the reduced prosthesis lifetime, compared to the esophageal speech patients ($p = 0.002$). However, the V-RQoL scoring showed a worsening of the well-being indices in the group with disorders related to the tracheo-esophageal fistula compared to the EV group ($p < 0.001$) (Table 3).

3.3. Voice Handicap Index (VHI) Score and Grading

The results of the patient groups analyzed are shown in the VHI score (Table 4). The TEP group showed better total VHI score than the EV group, but was not statistically significant (36.24 ± 7.19 vs. 38.53 ± 6.62; $p = 0.33$). Further examination of the pathological sub-phenotypes presented better scores in the voice prosthesis disorders group than in the esophageal speech (30.37 ± 4.88 vs. 38.53 ± 6.62; $p = 0.01$). However, the outcomes analysis of the patients with fistula disorders showed a worsening of the scores compared to esophageal speech (54.1 ± 10.48 vs. 38.53 ± 6.62; $p = 0.003$) (Table 4).

Table 4. Comparison of each sub-classes voice handicap index (VHI) scores. Statistical significance at $p < 0.05$.

	Tracheoesophageal Voice Prosthesis (TEP)	Voice Prosthesis Disorders (PD)	Fistula Related Disorders (FRD)	Esophageal Speech (EV)
VHI				
Emotional	9.59 ± 2.14	8.87 ± 0.99	13.1 ± 3.81	9.4 ± 1.35
Physical	12.12 ± 2.15	10.25 ± 1.58	18.5 ± 3.43	12.53 ± 2.58
Functional	14.53 ± 2.89	11.25 ± 2.31	22.5 ± 3.24	16.6 ± 2.69
Total Score	36.24 [a] ± 7.19	30.37 [b] ± 4.88	54.1 [c] ± 10.48	38.53 [a, b, c] ± 6.62

Abbreviations: Voice Related Quality of Life; TEP = tracheoesophageal prosthesis. Comparison: [a] TEP group vs. EV group $p = 0.33$; [b] PD group vs. EV group $p = 0.01$; [c] FRD vs. EV $p = 0.003$.

4. Discussion

The benefit of tracheoesophageal voice rehabilitation was formulated first in 1972 by Mozolewski et al. [14]. Since then, many devices have been produced by different companies with variable technologies and specific architecture features, generally taking advantage of a voice polymer prosthesis inserted through a puncture in the shared wall between the trachea and esophagus [11,15–19].

Countless clinical variables such as gastroesophageal reflux, ageing effect, adjuvant radiotherapy, or timing of surgery can influence the laryngectomee patient to failure of vocal rehabilitation treatment and, consequently, the quality of life (QoL) [7,20–25].

Precisely, pathological supraesophageal reflux correlates with the onset of fistula complications and the consecutive rehabilitation degree, inducing the onset of periprosthetic leakage [26]. In this regard, Lorenz et al. found higher VHI scores (up to 64.1 ± 9.6) with reflux severity ($p = 0.025$) and total quality of life scores were worse in patients with highly pathological reflux ($p = 0.007$).

Furthermore, the association of post-surgery radiotherapy in patients with voice prosthesis (PORT) can play a determining role in the genesis of gastroesophageal reflux and fistula-related pathology [27]. Cocuzza et al. compared two patient groups based on the choice of treatment and found a significantly higher rate of failure of voice rehabilitation in subjects with gastroesophageal reflux and history of postoperative radiotherapy (45% vs. 17%; $p < 0.05$).

The role of the correct therapeutic choice in the literature is much debated because of the possible consequent rehabilitation complications, making a careful selection of the patients' candidacy for TEP necessary [6,7,27–31]. For instance, the variable choice between radiotherapy protocols can lead to different long-term rehabilitative outcomes, depending on the therapeutic tissue dosage. As discussed by Elving et al., a dose equal or more than 60 Gray to the primary tumor site limited the prosthesis device lifetime ($p < 0.05$) [31].

The direct consequence consists of the remodeling of the pharyngeal microflora, which under physiological conditions produces mucins that are active against the primary pathogens of the Candida group and successive anomalous prosthetic colonization [32–34].

Later, Agarwal et al. discussed the role of treatment modalities including the surgical procedure and details such as neck dissection and pharyngeal reconstruction as well as radiation therapy on quality of life measured by the VHI and V-RQoL questionnaires [35]. Although the surgery did not show significant results, postoperative radiotherapy was initially associated with a higher level of voice handicap. However, the same score during follow-up was significantly decreased, according to the authors due to lessened tissue flexibility in early post-radiotherapy.

Either the disorders related to a reduced prosthetic lifetime or the fistula-related ones previously indicated can influence vocal and communication skills with significant sequelae on the quality of life, inducing the patients to make changes in their behavioral and social sphere [12,36].

The World Health Organization defines the quality of life as "the perception of the individual in his own life, in the context of cultural systems and values, of their objectives, expectations, standards and concerns" [5].

Undoubtedly, many determinants may limit tracheoesophageal voice rehabilitation, influencing either the patient's quality of life or vocal performance, but each of these can change the degree of treatment satisfaction variably. To this end, subjective questionnaires were designed to allow a self-assessment QoL in specific domains of an individual's life, also introducing the fundamental concept of voice-related quality of life [37–39].

Given these assumptions, Moukarbel et al. analyzed the voice-related quality of life (V-RQoL) outcomes specific to tracheoesophageal esophageal speech post-laryngectomy in 75 patients initially enrolled [39].

Although the paired comparison between the TEP and ES group showed a significant difference in socio-emotional and total scores ($p < 0.05$), however, the analysis of physical-functional domains was not significantly different. The data emerging from our analysis confirmed a statistical difference between the overall results of the TEP and EV groups (8.73 ± 4.71 vs. 10.76 ± 2.21; $p = 0.01$). Nevertheless, the observation of the two domains showed a significant result both in the functional $p = 0.01$ than in the socio-emotional one ($p = 0.01$).

Further assessment was performed by Agarwal et al., correlating voice-related QoL and socioeconomic status after total laryngectomy [35]. From the 104 patients who underwent total laryngectomy initially enrolled, only 71 were eligible for the study, administering the V-RQoL, and VHI questionnaires after 1-year of TEP rehabilitation. Long-term outcomes of the V-RQoL reported a higher patient satisfaction in the relation of the crucial contribution of social support (about 80% V-RQoL excellent score and >75% minimal VHI score).

In our study, although the TEP group patients obtained higher results than those in the esophageal speech group, statistical significance was not achieved when comparing VHI scores (36.24 ± 7.19 vs. 38.53 ± 6.62; $p = 0.33$). Therefore, the investigations on the relationship between comprehensibility, intensity, and what voice handicap can entail offers qualitative learning on the experience lived and the social impact of the voice prosthesis.

However, bias may be present in our findings from our non-randomized study regarding surgery type choice conducted, according to non-standardized protocols, and therefore outcomes may not be assessed blind.

Later in 2016, Ţiple et al. described the impact of vocal rehabilitation on quality of life and voice handicap in patients with a total laryngectomy, dealing in particular with the different adaptation timing to the rehabilitation method used [38].

Although the first use of the VHI questionnaire in patients with TEP was initially associated with severe voice handicap than the esophageal group (ES 52.67 ± 19.32 vs. TEP 61.57 ± 24.28); after an adjustment period of six months, the second application of the questionnaires showed a significant improvement in voice production (59.58 ± 16.33).

This assumption explains why patients with TEP need a longer considerable adaptation period to integrate back into society and suit the new conditions of life than esophageal voice.

In the previous study, our patient analysis based on the handicap of the voice after a follow-up of at least 10 years revealed a better, but not significant, TEP score of an EV, respectively (36.24 ± 7.19 vs. 38.53 ± 6.62; $p = 0.33$).

As described in the literature, the specific QoL outcomes of the TEP differ significantly according to the different risk factors that are concentrated within the TEP patients, therefore, the patients with excellent vocal performance, but reduced device lifetime, and patients with fistula related disorders [7–11,39,40]. Nevertheless, to our knowledge, this is the first published analysis of phenotyping different subgroups into tracheoesophageal speech patients.

In our study, a short device lifetime (<3 months) did not result in a limiting factor of tracheoesophageal (TE) voice restoration in both the V-RQoL score and the VHI compared with the EV group ($p = 0.002$; $p = 0.01$, respectively).

However, opposite data were found in the fistula-related disorders group with worse V-RQoL and VHI outcomes than patients with esophageal voice ($p = 0.0007$; $p = 0.003$, respectively).

As discussed above, despite favorable evidence regarding treatment, a percentage of subjects in the follow-up either considered removing the voice prosthesis or said that they would not choose the same type of voice rehabilitation if they could go back in time [10].

Limitations

The main limitation of the study we performed is the modest number of patients analyzed, which may have resulted in the research being underpowered. However, it is essential to consider that after applying such rigid inclusion and exclusion criteria such as an extended follow-up, the total number of selected patients was drastically reduced, making it a representative sample in any event.

Furthermore, bias may be present in our findings from our non-randomized study regarding surgery type choice conducted according to non-standardized protocols. Therefore, outcomes may not be assessed blind.

5. Conclusions

The prosthetic rehabilitation treatment allows for the recovery of the laryngectomee patient's communication skills, positively affecting both the cognitive-emotional component and the physical-functional one. The variable presence of sequelae acts as a risk factor for the worse quality of life of the patient, and the accurate choice of rehabilitative management represents the primary target. Although the TEP patient has excellent long-term outcomes, the correct phenotyping of complications can provide the keystone for better administration, identifying subjects who could better benefit from alternative rehabilitation procedures.

Author Contributions: Conceptualization, S.C., I.L.M., and A.M.; Methodology, A.M.; Validation, C.V., G.P.S., and S.C.; Formal analysis, C.V., G.I., S.C., and A.M.; Investigation, S.C., G.I., C.V., and C.G.; Resources, S.C., G.S.,

C.G., and I.L.M.; Data curation, G.I., F.M., S.C., I.L.M., and A.M.; Writing—original draft preparation, G.S., S.C., A.M., and S.F.; Writing—review and editing, G.S., S.C., A.M. All authors have read and agreed to the published version of the manuscript.

Funding: This research received no external funding.

Conflicts of Interest: The authors reported no potential conflicts of interest.

References

1. Ferlay, J.; Soerjomataram, I.; Dikshit, R.; Eser, S.; Mathers, C.; Rebelo, M.; Parkin, D.M.; Forman, D.; Bray, F. Cancer incidence and mortality worldwide: Sources, methods and major patterns in GLOBOCAN 2012. *Int. J. Cancer* **2015**, *136*, 359–386. [CrossRef]
2. Ferlay, J.; Colombet, M.; Soerjomataram, I.; Mathers, C.; Parkin, D.M.; Piñeros, M.; Znaor, A.; Bray, F. Estimating the global cancer incidence and mortality in 2018: GLOBOCAN sources and methods. *Int. J. Cancer* **2019**, *144*, 1941–1953. [CrossRef]
3. Saraniti, C.; Speciale, R.; Santangelo, M.; Massaro, N.; Maniaci, A.; Gallina, S.; Serra, A.; Cocuzza, S. Functional outcomes after supracricoid modified partial laryngectomy. *J. Biol. Regul. Homeost. Agents* **2019**, *33*, 1903–1907.
4. Serra, A.; Maiolino, L.; Di Mauro, P.; Licciardello, L.; Cocuzza, S. The senile functional evolution of the larynx after supracricoid reconstructive surgery. *Eur. Arch. Otorhinolaryngol.* **2016**, *273*, 4359–4368. [CrossRef] [PubMed]
5. Development of the World Health Organization WHOQOL-BREF quality of life assessment. The WHOQOL Group. *Psychol. Med.* **1998**, *28*, 551–558. [CrossRef] [PubMed]
6. Petersen, J.F.; Lansaat, L.; Timmermans, A.J.; van der Noort, V.; Hilgers, F.J.M.; van den Brekel, M.W.M. Postlaryngectomy prosthetic voice rehabilitation outcomes in a consecutive cohort of 232 patients over a 13-year period. *Head Neck* **2019**, *41*, 623–631. [CrossRef]
7. Serra, A.; Di Mauro, P.; Spataro, D.; Maiolino, L.; Cocuzza, S. Post-laryngectomy voice rehabilitation with voice prosthesis: 15 years experience of the ENT Clinic of University of Catania. Retrospective data analysis and literature review. *Acta Otorhinolaryngol. Ital.* **2015**, *35*, 412–419.
8. Souza, F.; Santos, I.C.; Bergmann, A.; Thuler, L.; Freitas, A.S.; Freitas, E.Q.; Dias, F.L. Quality of life after total laryngectomy: Impact of different vocal rehabilitation methods in a middle income country. *Health Qual Life Outcomes* **2020**, *18*, 92. [CrossRef]
9. Giordano, L.; Toma, S.; Teggi, R.; Palonta, F.; Ferrario, F.; Bondi, S.; Bussi, M. Satisfaction and quality of life in laryngectomees after voice prosthesis rehabilitation. *Folia Phoniatr. Logop.* **2011**, *63*, 231–236. [CrossRef]
10. Galli, A.; Giordano, L.; Biafora, M.; Tulli, M.; Di Santo, D.; Bussi, M. Voice prosthesis rehabilitation after total laryngectomy: Are satisfaction and quality of life maintained over time? *Acta Otorhinolaryngol. Ital.* **2019**, *39*, 162–168. [CrossRef]
11. Serra, A.; Spinato, G.; Spinato, R.; Conti, A.; Licciardello, L.; Di Luca, M.; Campione, G.; Tonoli, G.; Politi, D.; Castro, V.; et al. Multicenter prospective crossover study on new prosthetic opportunities in post-laryngectomy voice rehabilitation. *J. Biol. Regul. Homeost. Agents* **2017**, *31*, 803–809. [PubMed]
12. Sharpe, G.; Camoes Costa, V.; Doubé, W.; Sita, J.; McCarthy, C.; Carding, P. Communication changes with laryngectomy and impact on quality of life: A review. *Qual Life Res.* **2019**, *28*, 863–877. [CrossRef] [PubMed]
13. Sielska-Badurek, E.; Rzepakowska, A.; Sobol, M.; Osuch-Wójcikiewicz, E.; Niemczyk, K. Adaptation and Validation of the Voice-Related Quality of Life Measure Into Polish. *J. Voice* **2016**, *30*, 773.e7–773.e12. [CrossRef] [PubMed]
14. Mozolewski, E. Surgical rehabilitation of voice and speech following laryngectomy. *Otolaryngol. Pol.* **1972**, *26*, 653–661.
15. Hancock, K.; Houghton, B.; Van As-Brooks, C.J.; Coman, W. First clinical experience with a new non-indwelling voice prosthesis (Provox NID) for voice rehabilitation after total laryngectomy. *Acta Otolaryngol.* **2005**, *125*, 981–990. [CrossRef]
16. Hilgers, F.J.; Ackerstaff, A.H.; Jacobi, I.; Balm, A.J.; Tan, I.B.; van den Brekel, M.W. Prospective clinical phase II study of two new indwelling voice prostheses (Provox Vega 22.5 and 20 Fr) and a novel anterograde insertion device (Provox Smart Inserter). *Laryngoscope* **2010**, *120*, 1135–1143. [CrossRef]

17. Schuldt, T.; Ovari, A.; Dommerich, S. The costs for different voice prostheses depending on the lifetime. *Laryngorhinootologie* **2013**, *92*, 389–393.
18. Graville, D.; Gross, N.; Andersen, P.; Everts, E.; Cohen, J. The long-term indwelling tracheoesophageal prosthesis for alaryngeal voice rehabilitation. *Arch. Otolaryngol. Head Neck Surg.* **1999**, *125*, 288–292. [CrossRef]
19. Vlantis, A.C.; Gregor, R.T.; Elliot, H.; Oudes, M. Conversion from a non-indwelling to a Provox2 indwelling voice prosthesis for speech rehabilitation: Comparison of voice quality and patient preference. *J. Laryngol. Otol.* **2003**, *117*, 815–820. [CrossRef]
20. Chakravarty, P.D.; McMurran, A.E.L.; Banigo, A.; Shakeel, M.; Ah-See, K.W. Primary versus secondary tracheoesophageal puncture: Systematic review and meta-analysis. *J. Laryngol. Otol.* **2018**, *132*, 14–21. [CrossRef]
21. Gazia, F.; Galletti, B.; Freni, F.; Bruno, R.; Sireci, F.; Galletti, C.; Meduri, A.; Galletti, F. Use of intralesional cidofovir in the recurrent respiratory papillomatosis: A review of the literature. *Eur. Rev. Med. Pharmacol. Sci.* **2020**, *24*, 956–962. [PubMed]
22. Ferlito, S.; Di Luca, S.; Maniaci, A.; Cipolla, F.; La Mantia, I.; Maiolino, L.; Ferlito, A.; Cocuzza, S.; Grillo, C. Progressive dysphagia in a patient with parapharingeal pulsating mass: A case report and literature's review. *Acta Med. Mediterr.* **2019**, *35*, 3433.
23. Galletti, F.; Freni, F.; Gazia, F.; Gallo, A. Vocal cord surgery and pharmacological treatment of a patient with HPV and recurrent respiratory papillomatosis. *BMJ Case. Rep.* **2019**, *12*, e231117. [CrossRef] [PubMed]
24. Cocuzza, S.; Bonfiglio, M.; Chiaramonte, R.; Aprile, G.; Mistretta, A.; Grosso, G.; Serra, A. Gastroesophageal reflux disease and postlaryngectomy tracheoesophageal fistula. *Eur. Arch. Otorhinolaryngol.* **2012**, *269*, 1483–1488. [CrossRef]
25. Cocuzza, S.; Bonfiglio, M.; Grillo, C.; Maiolino, L.; Malaguarnera, M.; Martines, F.; Serra, A. Post laryngectomy speech rehabilitation outcome in elderly patients. *Eur. Arch. Otorhinolaryngol.* **2013**, *270*, 1879–1884. [CrossRef]
26. Lorenz, K.J.; Grieser, L.; Ehrhart, T.; Maier, H. Laryngectomised patients with voice prostheses: Influence of supra-esophageal reflux on voice quality and quality of life. *HNO* **2011**, *59*, 179–187. [CrossRef]
27. Cocuzza, S.; Bonfiglio, M.; Chiaramonte, R.; Serra, A. Relationship between radiotherapy and gastroesophageal reflux disease in causing tracheoesophageal voice rehabilitation failure. *J. Voice* **2014**, *28*, 245–249. [CrossRef]
28. Cocuzza, S.; Di Luca, M.; Maniaci, A.; Russo, M.; Di Mauro, P.; Migliore, M.; Serra, A.; Spinato, G. Precision treatment of post pneumonectomy unilateral laryngeal paralysis due to cancer. *Future Oncol.* **2020**, *16*, 45–53. [CrossRef]
29. Galletti, B.; Costanzo, D.; Gazia, F.; Galletti, F. High-grade chondrosarcoma of the larynx: Treatment and management. *BMJ Case Rep.* **2019**, *12*, e230918. [CrossRef]
30. Kummer, P.; Chahoud, M.; Schuster, M.; Eysholdt, U.; Rosanowski, F. Prosthetic voice rehabilitation after laryngectomy. Failures and complications after previous radiation therapy. *HNO* **2006**, *54*, 315–322. [CrossRef]
31. Elving, G.J.; Van Weissenbruch, R.; Busscher, H.J.; Van Der Mei, H.C.; Albers, F.W. The influence of radiotherapy on the lifetime of silicone rubber voice prostheses in laryngectomized patients. *Laryngoscope* **2002**, *112*, 1680–1683. [CrossRef] [PubMed]
32. Talpaert, M.J.; Balfour, A.; Stevens, S.; Baker, M.; Muhlschlegel, F.A.; Gourlay, C.W. Candida biofilm formation on voice prostheses. *J. Med. Microbiol.* **2015**, *64*, 199–208. [CrossRef] [PubMed]
33. Bauters, T.G.; Moerman, M.; Vermeersch, H.; Nelis, H.J. Colonization of voice prostheses by albicans and non-albicans Candida species. *Laryngoscope* **2002**, *112*, 708–712. [CrossRef] [PubMed]
34. Grillo, C.; Saita, V.; Grillo, C.M.; Andaloro, C.; Oliveri, S.; Trovato, L.; La Mantia, I. Candida colonization of silicone voice prostheses: Evaluation of device lifespan in laryngectomized patients. *Otorhinolaryngology* **2017**, *67*, 75–80.
35. Agarwal, S.K.; Gogia, S.; Agarwal, A.; Agarwal, R.; Mathur, A.S. Assessment of voice related quality of life and its correlation with socioeconomic status after total laryngectomy. *Ann. Palliat. Med.* **2015**, *4*, 169–175. [PubMed]
36. Summers, L. Social and quality of life impact using a voice prosthesis after laryngectomy. *Curr. Opin. Otolaryngol. Head Neck Surg.* **2017**, *25*, 188–194. [CrossRef]

37. Hogikyan, N.; Sethuraman, G. Validation of an instrument to measure voice-related quality of life (V-RQOL). *J. Voice* **1999**, *13*, 557–569. [CrossRef]
38. Ţiple, C.; Drugan, T.; Dinescu, F.V.; Mureşan, R.; Chirilă, M.; Cosgarea, M. The impact of vocal rehabilitation on quality of life and voice handicap in patients with total laryngectomy. *J. Res. Med. Sci.* **2016**, *21*, 127.
39. Moukarbel, R.V.; Doyle, P.C.; Yoo, J.H.; Franklin, J.H.; Day, A.M.; Fung, K. Voice-related quality of life (V-RQOL) outcomes in laryngectomees. *Head Neck* **2011**, *33*, 31–36. [CrossRef]
40. Lewin, J.S.; Baumgart, L.M.; Barrow, M.P.; Hutcheson, K.A. Device Life of the Tracheoesophageal Voice Prosthesis Revisited. *JAMA Otolaryngol. Head Neck Surg.* **2017**, *143*, 65–71. [CrossRef]

International Journal of
Environmental Research and Public Health

Article

Association between Anemia and Auditory Threshold Shifts in the US Population: National Health and Nutrition Examination Survey

Jui-Hu Shih [1,2], I-Hsun Li [1,2,3], Ke-Ting Pan [4,5], Chih-Hung Wang [6,7], Hsin-Chien Chen [8], Li-Yun Fann [9,10], Jen-Ho Tseng [11] and Li-Ting Kao [1,2,12,13,*]

1 Department of Pharmacy Practice, Tri-Service General Hospital, Taipei 11490, Taiwan; jtlovehl@gmail.com (J.-H.S.); lhs01077@gmail.com (I.-H.L.)
2 School of Pharmacy, National Defense Medical Center, Taipei 11490, Taiwan
3 Department of Pharmacology, National Defense Medical Center, Taipei 11490, Taiwan
4 Institute of Environmental Design and Engineering, Bartlett School, University College London, London WC1E, UK; pankt930@gmail.com
5 Graduate Institute of Aerospace and Undersea Medicine, National Defense Medical Center, Taipei 11490, Taiwan
6 Graduate Institute of Medical Sciences, National Defense Medical Center, Taipei 11490, Taiwan; chw@ms3.hinet.net
7 Department of Otorhinolaryngology, Taichung Armed Forces General Hospital, Taichung 41152, Taiwan
8 Department of Otorhinolaryngology-Head and Neck Surgery, Tri-Service General Hospital, Taipei 11490, Taiwan; acolufreia@yahoo.com.tw
9 Department of Nursing, Taipei City Hospital, Taipei 106, Taiwan; fanly99@gmail.com
10 Department of Biology and Anatomy, National Defense Medical Center, Taipei 11490, Taiwan
11 Department of Neurosurgery, Taipei City Hospital, Taipei 106, Taiwan; DAL87@tpech.gov.tw
12 Graduate Institute of Life Sciences, National Defense Medical Center, Taipei 11490, Taiwan
13 School of Public Health, National Defense Medical Center, Taipei 11490, Taiwan
* Correspondence: kaoliting@gmail.com; Tel.: +886-2-8792-3311 (ext. 17629); Fax: +886-2-8792-3171

Received: 22 April 2020; Accepted: 28 May 2020; Published: 1 June 2020

Abstract: Existing evidence indicates that both iron deficiency anemia and sickle cell anemia have been previously associated with hearing loss. However, human data investigating the association between anemia and auditory threshold shifts at different frequencies in the adolescent, adult and elderly population are extremely limited to date. Therefore, this cross-sectional study used the dataset from the US National Health and Nutrition Examination Survey from 2005 to 2012 to explore differences in low- or high-frequency hearing thresholds and hearing loss prevalence between participants with and without anemia. A total of 918 patients with anemia and 8213 without anemia were included. Results indicated that low- and high-frequency pure tone average were significantly higher in patients with anemia than that in those without anemia in the elderly, but not in adult or adolescent population. In addition, the prevalence of low-frequency hearing loss but not high-frequency hearing loss was also higher in patients with anemia than in those without anemia in the elderly population. After adjusting various confounders, multiple regression models still indicated that patients with anemia tended to have larger threshold shift. In conclusion, anemia was associated with auditory threshold shifts in the elderly population, especially those vulnerable to low-frequency hearing loss.

Keywords: anemia; hearing loss; auditory threshold shifts; pure tone average

1. Introduction

Approximately 466 million people worldwide have disabling hearing loss in 2018 [1], and unaddressed hearing loss poses an annual global cost of US$ 750–790 billion [2]. In the United States, the number of adults with hearing loss (pure tone average (PTA), >25 dB) is estimated to gradually increase from 44.1 million in 2020 to 73.5 million by 2060 [3]. Hearing loss is a sensory impairment caused by multiple factors including gene mutations, environmental noise, viral infection, autoimmune disease, labyrinthine membrane rupture, vascular events (e.g., vascular disease/alteration of microcirculation, vascular disease associated with mitochondriopathy, vertebrobasilar insufficiency, red blood cell deformability, sickle cell disease and cardiopulmonary bypass), and blood disorders [4]. Anemia is the most common blood disorder and also remains the major global public health problems. Existing evidence indicated that both iron deficiency anemia (IDA) and sickle cell anemia (SCA) have been previously associated with hearing loss [5,6].

In a Taiwan population-based study, Chung et al. found that the odds ratio (OR) for having a previous IDA diagnosis among patients aged ≥18 years with sudden sensorineural hearing loss was 1.34 (95% confidence interval (CI), 1.11–1.61), which was most pronounced among those aged ≤44 years compared with controls (OR, 1.91; 95% CI, 1.35–2.72) [7]. US cohort studies by Schieffer et al. also showed that children and adolescents with IDA have an increased risk of developing sensorineural hearing loss (OR, 3.67; 95% CI, 1.60–7.30); increased odds of sensorineural hearing loss (OR, 1.82; 95% CI, 1.18–2.66) but conductive hearing loss (OR, 1.51; 95% CI, 0.54–3.28) among adults with IDA [8,9]. In addition, a prospective case-control study by Aderibigbe et al. indicated that the average hearing thresholds of patients with SCA aged 16–48 years were significantly higher than controls aged 15–39 years in both right and left ears (right hearing thresholds: 16.7 ± 13.0 vs. 11.1 ± 9.1 dB; left hearing thresholds: 12.8 ± 9.4 vs. 9.6 ± 5.8 dB) [10].

Although IDA and SCA mechanisms that lead to hearing loss differ, hemoglobin may play a common and critical role in the deterioration of hearing function. In addition, typical changes in age-related hearing loss usually start with a hearing loss on high frequencies [11], and there are rare reports of low-frequency hearing loss occurring in old age. To date, human data investigating the association between anemia and auditory threshold shifts at different frequencies in the adolescent, adult and elderly population are extremely limited. Therefore, we use the dataset from the National Health and Nutrition Examination Survey (NHANES) to explore differences in low- or high-frequency hearing thresholds and hearing loss prevalence between participants with and without anemia.

2. Materials and Methods

2.1. Database

The data used in this cross-sectional study were derived from the NHANES in the United States (https://wwwn.cdc.gov/nchs/nhanes/). This survey includes information about questionnaires, demographic data, laboratory tests, and physical examinations, among others. All the study participants in the NHANES are sampled from the residents in the United States. Its research protocols were approved by the Research Ethics Review Board of National Center for Health Statistics. All participants provided written informed consents. Because we only used de-identified secondary data from NHANES, this study was exempted from full review by the Institutional Review Board.

2.2. Study Participants Selection and Anemia Definition

This study attempted to identify the association between anemia and auditory threshold shifts. Therefore, study participants aged 12 years and older were only limited to 2005–2012 NHANES individuals who underwent both audiometric examinations and with complete blood count results. Those with incomplete audiometric exams or missing laboratory tests for hemoglobin were excluded from this study. A total of 9131 US participants were recruited in this cross-sectional study. We then classified the study participants into two groups: patients with and without anemia. Anemia was defined based on the World Health Organization recommendations. Male participants with the serum hemoglobin level of <13 g/dL or female participants with the serum hemoglobin level of <12 g/dL were identified as patients with anemia [12]. Finally, 918 patients with anemia and 8213 without anemia were included in this study.

2.3. Audiometric Measures

According to the NHANES protocols, audiometric examinations excluded participants who could not tolerate headphones because of ear pain at the exam time. Those using hearing aids (not able to remove them) during the test were also excluded. All audiometry examinations were carried out by professionally trained audiologists from the National Institute for Occupational Safety & Health using the interacoustics model AD226 audiometer (with standard TDH-49 headphones and Etymotic EarTone 3A insert earphones). The hearing threshold test in this study was conducted on both right and left ears at 0.5, 1, 2, 3, 4, 6, and 8 kHz across an intensity range of −10 to 120 decibels (dB). Full audiometric protocols and procedures are available online.

2.4. Auditory Threshold Shifts and Hearing Loss Definition

In the relevant analyses of this study, PTA thresholds of hearing at 0.5, 1, and 2 kHz were identified as the low-frequency PTA (low PTA) and PTA thresholds of hearing at 3, 4, and 6 kHz were identified as the high-frequency PTA (high PTA) [13]. Additionally, this cross-sectional study identified patients with PTA at 0.5, 1, and 2 kHz thresholds of ≥15 dB in either ear as those with low-frequency hearing loss (LFHL). Patients who met the criteria for PTA at 3, 4, and 6 kHz thresholds of ≥15 dB in whichever ear were defined as the high-frequency hearing loss (HFHL). Furthermore, patients who experienced any hearing loss of ≥15 dB (including HFHL or LFHL) were identified as those with overall hearing loss (HL).

2.5. Covariate Measurement

To eliminate the potential effects of some confounders and investigate the actual relationship between anemia and HL, participants' age group, sex, race, hypertension, diabetes, coronary heart disease, heart failure, and stroke were considered in the regression models. In this study, the race was categorized as non-Hispanic white, non-Hispanic black, or others (including Mexican American, other Hispanic, and other races). In addition, participants were considered to have medical history of hypertension, coronary heart disease, heart failure, and stroke if participants self-reported to have medical history about these diseases according to their physicians. Moreover, patients self-reported to have diabetes by physicians or other health professionals or receive diabetic medications were identified as diabetes cases.

2.6. Statistical Analysis

All analyses were performed with the SAS system (SAS System for Windows, ver. 9.4, SAS Institute Inc., Cary, NC, USA). This study used the chi-squared tests to explore differences in age group, sex, race, hypertension, diabetes, coronary heart disease, heart failure, and stroke, LFHL, HFHL, HL, between participants with and without anemia. Independent t-tests were further conducted to compare differences in low PTA and high PTA between participants with and without anemia. Additionally, regression models were carried out to estimate the effects of anemia on PTA thresholds of hearing. A two-sided p-value of 0.05 was used to determine the statistical significance of this study.

3. Results

This study consisted of 918 participants with anemia and 8213 without anemia. Table 1 shows the demographic characteristics and comorbidities of participants with anemia and controls without anemia. Relevant findings indicated that there were significant differences in age group ($p < 0.001$), sex ($p < 0.001$), and race ($p < 0.001$) between the two groups. In addition, the prevalence of hypertension ($p < 0.001$), diabetes ($p < 0.001$), coronary heart disease ($p < 0.001$), heart failure ($p < 0.001$), and stroke ($p < 0.001$) between participants with anemia and controls without anemia was significantly different.

Table 1. Demographic characteristics and comorbidities of subjects with anemia and the controls without anemia (n = 9131).

Variable	Subjects with Anemia (n = 918)		Subjects without Anemia (n = 8213)		p Value
	No.	%	No.	%	
Age Group (Years)					**<0.001**
≤19	313	34.1	3675	44.8	
20–29	72	7.8	721	8.8	
30–39	73	8.0	664	8.1	
40–49	85	9.3	629	7.7	
50–59	68	7.4	673	8.2	
60–69	74	8.1	610	7.4	
≥70	233	25.4	1241	15.1	
Sex					<0.001
Male	367	40.0	4292	52.3	
Female	551	60.0	3921	47.7	
Race					<0.001
Non-Hispanic white	224	24.4	3285	40.0	
Non-Hispanic black	446	48.6	1871	22.8	
Other	248	27.0	3057	37.2	
Ever had diagnosis					
Hypertension	329	35.8	1835	22.3	<0.001
Diabetes	128	13.9	570	6.9	<0.001
Coronary heart disease	44	4.8	212	2.6	<0.001
Heart failure	42	4.6	154	1.9	<0.001
Stroke	49	5.3	178	2.2	<0.001

Table 2 displays the high PTA, low PTA, and prevalence of LFHL, HFHL, and HL in participants with and without anemia. Results showed that participants with anemia had significantly higher low PTA (right ear: 15.04 dB vs. 11.64 dB; left ear: 15.57 dB vs. 11.56 dB) and high PTA (right ear: 23.82 dB vs. 18.55 dB; left ear: 24.71 dB vs. 19.51 dB) in both ears compared to participants without anemia. Moreover, the HL was found in 64.16% of participants with anemia and in 54.60% of controls without anemia ($p < 0.001$). LFHL was observed in 42.59% of participants with anemia and in 32.66% of controls without anemia ($p < 0.001$). Furthermore, the HFHL was found in 61.44% of participants with anemia and in 51.88% of controls without anemia ($p < 0.001$).

Table 2. High-PTA, low-PTA, and prevalence of hearing loss in subjects with and without anemia.

Variables	Total (n = 9131)		Subjects with Anemia (n = 918)		Subjects without Anemia (n = 8213)		p Value
	n, %		n, %		n, %		
Low-PTA (dB)							
Right Ear	11.98 ± 12.58		15.04 ± 14.45		11.64 ± 12.31		<0.001
Left Ear	11.97 ± 12.71		15.57 ± 15.37		11.56 ± 12.31		<0.001
High-PTA (dB)							
Right Ear	19.08 ± 20.74		23.82 ± 23.06		18.55 ± 20.39		<0.001
Left Ear	20.04 ± 21.35		24.71 ± 23.80		19.51 ± 20.99		<0.001
Hearing Loss							
HL	5073	55.56	589	64.16	4484	54.60	<0.001
LFHL	3073	33.65	391	42.59	2682	32.66	<0.001
HFHL	4825	52.84	564	61.44	4261	51.88	<0.001

Low-PTA, pure tone average at low frequencies; High-PTA, pure tone average at high frequencies; LFHL, low-frequency hearing loss; HFHL, high-frequency hearing loss; HL, hearing loss.

Because age is an important factor in the development of HL, Table 3 further displays the high PTA, low PTA, and prevalence of LFHL, HFHL, HL in participants with and without anemia according to the age group. Independent t-tests presented that participants with anemia had a significantly greater low PTA and high PTA than those without anemia among the elderly population (aged ≥ 60 years) in both ears. The prevalence of LFHL was also higher in participants with anemia than that in those without anemia. However, no significant difference in low PTA and high PTA was observed between patients with and without anemia among adult and adolescent population.

Table 3. High-PTA, low-PTA, and prevalence of hearing loss in subjects with and without anemia according to the age group.

	Elderly Population (≥60 Years Old)			Adult Population (20–59 Years Old)			Adolescent Population (≤19 Years Old)		
Variables	Subjects with Anemia (n = 307)	Subjects without Anemia (n = 1851)	p Value	Subjects with Anemia (n = 298)	Subjects without Anemia (n = 2687)	p Value	Subjects with Anemia (n = 313)	Subjects without Anemia (n = 3675)	p Value
Low-PTA (dB)									
Right Ear	28.07 ± 15.23	25.13 ± 15.32	0.002	9.93 ± 8.98	9.24 ± 9.15	0.213	7.14 ± 7.41	6.60 ± 6.15	0.212
Left Ear	29.25 ± 16.49	25.01 ± 15.30	<0.001	10.55 ± 10.09	9.21 ± 8.84	0.028	6.93 ± 6.90	6.51 ± 6.57	0.278
High-PTA (dB)									
Right Ear	48.20 ± 20.40	45.22 ± 21.57	0.024	15.44 ± 13.56	15.82 ± 14.30	0.231	7.88 ± 8.55	7.12 ± 7.16	0.129
Left Ear	50.54 ± 20.50	47.00 ± 21.52	0.007	15.79 ± 13.05	16.98 ± 14.85	0.140	7.87 ± 8.57	7.52 ± 7.83	0.490
Hearing Loss									
HL	306 (99.67)	1825 (98.60)	0.163	185 (62.08)	1655 (61.59)	0.870	98 (31.31)	1004 (27.32)	0.130
LFHL	268 (87.30)	1494 (80.71)	0.006	80 (26.85)	717 (26.68)	0.952	43 (13.74)	471 (12.82)	0.640
HFHL	305 (99.35)	1820 (98.33)	0.216	178 (59.73)	1606 (59.77)	0.990	81 (25.88)	835 (22.72)	0.202

Note: Low-PTA, pure tone average at low frequencies; High-PTA, pure tone average at high frequencies; LFHL, low-frequency hearing loss; HFHL, high-frequency hearing loss; HL, hearing loss.

To reduce the potential effects of confounders, the regression models were used to investigate the relationships between PTA and anemia according to different age groups (Table 4). After adjusting various confounders, multiple regression models still indicated that patients with anemia tended to have higher hearing thresholds. Anemia was also significantly positively associated with the hearing thresholds, including both high PTA and low PTA, in the overall and elderly population. β coefficients of the high PTA comparing participants with anemia to controls without anemia were 2.35 for the right ear ($p < 0.001$) and 2.33 for the left ear ($p < 0.001$), and β coefficients of the low PTA were 1.53 for the right ear ($p < 0.001$) and 2.25 for the left ear ($p < 0.001$) in the overall population. Additionally, β coefficients of the high PTA comparing participants with anemia to controls without anemia were 3.76 for the right ear ($p = 0.002$), and 4.03 for the left ear ($p = 0.001$) and β coefficients of the low PTA were 3.55 for the right ear ($p < 0.001$) and 4.64 for the left ear ($p < 0.001$) in the elderly population. Additionally, this study performed the regression models to investigate the relationships between PTA and anemia according to different sexes and ethnic groups (Supplementary Table S1). After adjustments, anemia was positively associated with the hearing thresholds, including both high PTA and low PTA, in women, men, non-Hispanic white, and non-Hispanic black populations.

Table 4. Regression analyses of relationships between pure tone average and anemia according to different age groups.

Models	Variables	High-PTA (dB)				Low-PTA (dB)			
		Right Ear		Left Ear		Right Ear		Left Ear	
		β	*p* Value	β	*p* Value	β	*p* Value	β	*p* Value
Model 1 [a]	All Patients with Anemia	5.27	<0.001	5.20	<0.001	3.40	<0.001	4.01	<0.001
	Elderly Patients with Anemia	2.99	0.024	3.54	0.007	2.93	0.002	4.24	<0.001
	Adult Patients with Anemia	−0.38	0.662	−1.20	0.182	0.69	0.213	1.34	0.015
	Adolescent Patients with Anemia	0.76	0.077	0.35	0.455	0.54	0.144	0.42	0.278
Model 2 [b]	All Patients with Anemia [c]	2.35	<0.001	2.33	<0.001	1.53	<0.001	2.25	<0.001
	Elderly Patients with Anemia [d]	3.76	0.002	4.03	0.001	3.55	<0.001	4.64	<0.001
	Adult Patients with Anemia [d]	1.42	0.102	1.42	0.109	0.55	0.340	1.37	0.015
	Adolescent Patients with Anemia [e]	0.79	0.073	0.59	0.210	0.65	0.085	0.77	0.053

Note: Low-PTA, pure tone average at low frequencies; High-PTA, pure tone average at high frequencies.[a] Model 1: Univariate regression; [b] Model 2: Multiple regression; [c] Adjusted covariates: sex, race, age, hypertension, diabetes, coronary heart disease, heart failure, stroke; [d] Adjusted covariates: sex, race, hypertension, diabetes, coronary heart disease, heart failure, stroke; [e] Adjusted covariates: sex and race.

4. Discussion

In this study, anemia was found to be associated with auditory threshold shifts, which can be used to explain the higher prevalence of LFHL in elderly participants with anemia. To date, the mechanism of anemia-associated HL remains unclear and non-conclusive, although some plausible mechanisms were proposed. In the cochlear, intricate vasculature provides oxygen and nutrients needed for the stria vascularis of the cochlear duct to maintain the ionic composition of the endolymph and the endocochlear potential [14,15]. As anemia could decrease oxygen delivery in the labyrinthine arterial blood due to reduced hemoglobin concentration, blood oxygen supply to the cochlear is highly sensitive to ischemic damage.

Previous studies indicated that IDA is a potential risk factor for ischemic stroke [16], and patients with vascular disease have a higher risk for developing sudden sensorineural HL [7,17–20]. Iron is not only an essential component of hemoglobin in the red blood cells for tissue oxygen delivery but also a cofactor in neurotransmitter metabolism, DNA synthesis, and nerve myelination [21–24]. Neurological disorders might be associated with sensorineural HL. A systematic review and meta-analysis showed that compared with individuals without IDA, the age-specific OR of sensorineural HL was higher for children (3.67, 95% CI 1.72–7.84) than for adults (1.36, 95% CI 1.15–1.61; $p = 0.27$) [5]. However, participants with anemia have higher values of low and high PTA than those without anemia in the elderly but not the adult or adolescent population in this study (Table 3). We found that adolescent population with anemia are not apparent with low PTA threshold shift of 0.42–0.54 dB and high PTA threshold shift of 0.35–0.76 dB; low PTA threshold shift of 0.69–1.34 dB but high PTA negative threshold shift of 0.38–1.19 dB in the adult population. Elderly participants with anemia have remarkably low PTA threshold shift of 2.94–4.24 dB and high PTA threshold shift of 2.98–3.54 dB. Therefore, it is speculated that the biological interaction of aging and anemia would be a chronic progressive contributing factor to HL in the elderly population. In contrast, it may be different from IDA that has been regarded as a deteriorating factor of sudden sensorineural HL among adults aged <44 years [7].

Presbycusis, or age-related HL, is a common disorder characterized by symmetrical progressive loss of high-frequency hearing over the years. The World Health Organization estimates that >500 million people aged >60 years worldwide will suffer significant impairment from presbycusis by 2025 [25]. In this study, although elderly people with anemia have higher high PTA threshold values, similar prevalence of HFHL was found between the elderly people with (99.35%) and without anemia (98.33%). It is noteworthy that anemia was associated with LFHL only in the elderly population. The prevalence of LFHL in elderly people with anemia is 87.30%, which is higher than 80.71% in those without anemia. Based on our study results, we should particularly check and correct the hemoglobin level of elderly patients in routine clinical care to prevent the chronic progressive development of LFHL.

Limitations

This study had several limitations. First, all data in NHANES are cross-sectional, and thus this study cannot establish the causal relationship between anemia and hearing threshold shift. Second, we cannot assess environmental, occupational, or recreational noises around the participants included in this study. Previous reports have highlighted the significance of noise-induced HL from both work and recreational activities [26,27]. Third, this study also could not evaluate the use of medications among these participants. Some drugs, such as gentamicin, sildenafil, and cisplatin chemotherapy, can damage the inner ear [28–30]. In addition, high doses of aspirin, other non-steroidal anti-inflammatory drugs, antimalarial drugs, or loop diuretics have also been reported to cause temporary tinnitus or HL [31]. Fourth, information about ear infection and abnormal bone growths or tumors in the outer or middle ear that can also result in HL was not available in the NHANES. Fifth, LFHL may be caused by Meniere's disease, genetic conditions, central lesions, low spinal fluid pressure, and lithium use. These confounders cannot be adjusted in the multiple regression analysis of this study. Sixth, LFHL is also often related to conductive hearing loss, but bone thresholds are not available from the NHANES dataset. Therefore, we could not exclude conductive LFHL in this study. Seventh, previous evidence indicated that both anemia and hearing loss are sometimes partly explained by genetic deviations. Genetic components of anemia and HL were not available for this study. In addition, we did not consider some chronic diseases, like kidney diseases, rheumatism and cancer, which could be the comorbidity of both anemia and HL. Finally, the study findings should be cautiously generalized to other ethnicities because all patients included in this study were US residents.

In conclusion, our study demonstrated that anemia was associated with LFHL in the elderly population. However, further larger epidemiologic studies need to be conducted to confirm the effects of anemia on auditory threshold shifts in different ethnic groups and countries.

Supplementary Materials: The following are available online at http://www.mdpi.com/1660-4601/17/11/3916/s1, Table S1: Regression analyses of relationships between pure tone average and anemia according to sex and ethnicities.

Author Contributions: L.-T.K. had full access to all of the data in the study and take responsibility for the integrity of the data and the accuracy of the data analysis. Concept and design: J.-H.S., I.-H.L., and L.-T.K. Acquisition, analysis, or interpretation of data: J.-H.S. and L.-T.K. Drafting of the manuscript: J.-H.S. and L.-T.K. Critical revision of the manuscript for important intellectual content: J.-H.S., L.-T.K., I.-H.L., C.-H.W., H.-C.C., L.-Y.F. and J.-H.T. Statistical analysis: K.-T.P. and L.-T.K. All authors have read and agreed to the published version of the manuscript.

Funding: This work was supported by Tri-Service General Hospital, Taipei, Taiwan (TSGH-D-109145 and TSGH-E-109242) and Department of Health, Taipei City Government, Taipei, Taiwan (10801-62-072).

Conflicts of Interest: The authors declare no conflict of interest.

References

1. WHO. *Addressing the Rising Prevalence of Hearing Loss*; World Health Organization: Geneva, Switzerland, 2018.
2. WHO. *Global Costs of Unaddressed Hearing Loss and Cost-Effectiveness of Interventions: A Who Report, 2017*; World Health Organization: Geneva, Switzerland, 2017.
3. Goman, A.M.; Reed, N.S.; Lin, F.R. Addressing estimated hearing loss in adults in 2060. *JAMA Otolaryngol. Head Neck Surg.* **2017**, *143*, 733–734. [CrossRef]
4. Rizzo, S.; Bentivegna, D.; Thomas, E.; La Mattina, E.; Mucia, M.; Salvago, P.; Sireci, F.; Martines, F. *Sudden Sensorineural Hearing Loss, an Invisible Male: State of the Art*; Nova Science Publishers, Inc.: New York, NY, USA, 2017.
5. Mohammed, S.H.; Shab-Bidar, S.; Abuzerr, S.; Habtewold, T.D.; Alizadeh, S.; Djafarian, K. Association of anemia with sensorineural hearing loss: A systematic review and meta-analysis. *BMC Res. Notes* **2019**, *12*, 283. [CrossRef]
6. Lago, M.R.R.; Fernandes, L.D.C.; Lyra, I.M.; Ramos, R.T.; Teixeira, R.; Salles, C.; Ladeia, A.M.T. Sensorineural hearing loss in children with sickle cell anemia and its association with endothelial dysfunction. *Hematology* **2018**, *23*, 849–855. [CrossRef]

7. Chung, S.D.; Chen, P.Y.; Lin, H.C.; Hung, S.H. Sudden sensorineural hearing loss associated with iron-deficiency anemia: A population-based study. *JAMA Otolaryngol. Head Neck Surg.* **2014**, *140*, 417–422. [CrossRef]
8. Schieffer, K.M.; Chuang, C.H.; Connor, J.; Pawelczyk, J.A.; Sekhar, D.L. Association of iron deficiency anemia with hearing loss in us adults. *JAMA Otolaryngol. Head Neck Surg.* **2017**, *143*, 350–354. [CrossRef]
9. Schieffer, K.M.; Connor, J.R.; Pawelczyk, J.A.; Sekhar, D.L. The relationship between iron deficiency anemia and sensorineural hearing loss in the pediatric and adolescent population. *Am. J. Audiol.* **2017**, *26*, 155–162. [CrossRef]
10. Aderibigbe, A.; Ologe, F.E.; Oyejola, B.A. Hearing thresholds in sickle cell anemia patients: Emerging new trends? *J. Natl. Med. Assoc.* **2005**, *97*, 1135–1142.
11. Gates, G.A.; Mills, J.H. Presbycusis. *Lancet* **2005**, *366*, 1111–1120. [CrossRef]
12. Beutler, E.; Waalen, J. The definition of anemia: What is the lower limit of normal of the blood hemoglobin concentration? *Blood* **2006**, *107*, 1747–1750. [CrossRef]
13. Su, B.M.; Chan, D.K. Prevalence of hearing loss in us children and adolescents: Findings from nhanes 1988–2010. *JAMA Otolaryngol. Head Neck Surg.* **2017**, *143*, 920–927. [CrossRef]
14. Lamm, K.; Arnold, W. The effect of blood flow promoting drugs on cochlear blood flow, perilymphatic po(2) and auditory function in the normal and noise-damaged hypoxic and ischemic guinea pig inner ear. *Hear Res.* **2000**, *141*, 199–219. [CrossRef]
15. Shi, X. Physiopathology of the cochlear microcirculation. *Hear Res.* **2011**, *282*, 10–24. [CrossRef]
16. Chang, Y.L.; Hung, S.H.; Ling, W.; Lin, H.C.; Li, H.C.; Chung, S.D. Association between ischemic stroke and iron-deficiency anemia: A population-based study. *PLoS ONE* **2013**, *8*, e82952. [CrossRef]
17. Capaccio, P.; Ottaviani, F.; Cuccarini, V.; Bottero, A.; Schindler, A.; Cesana, B.M.; Censuales, S.; Pignataro, L. Genetic and acquired prothrombotic risk factors and sudden hearing loss. *Laryngoscope* **2007**, *117*, 547–551. [CrossRef]
18. Marcucci, R.; Alessandrello Liotta, A.; Cellai, A.P.; Rogolino, A.; Berloco, P.; Leprini, E.; Pagnini, P.; Abbate, R.; Prisco, D. Cardiovascular and thrombophilic risk factors for idiopathic sudden sensorineural hearing loss. *J. Thromb. Haemost.* **2005**, *3*, 929–934. [CrossRef]
19. Rudack, C.; Langer, C.; Stoll, W.; Rust, S.; Walter, M. Vascular risk factors in sudden hearing loss. *Thromb. Haemost.* **2006**, *95*, 454–461. [CrossRef] [PubMed]
20. Chung, J.H.; Lee, S.H.; Park, C.W.; Kim, C.; Park, J.K.; Shin, J.H. Clinical significance of arterial stiffness in idiopathic sudden sensorineural hearing loss. *Laryngoscope* **2016**, *126*, 1918–1922. [CrossRef] [PubMed]
21. Lopez, A.; Cacoub, P.; Macdougall, I.C.; Peyrin-Biroulet, L. Iron deficiency anaemia. *Lancet* **2016**, *387*, 907–916. [CrossRef]
22. Camaschella, C. Iron-deficiency anemia. *N. Engl. J. Med.* **2015**, *372*, 1832–1843. [CrossRef]
23. Zhang, C. Essential functions of iron-requiring proteins in DNA replication, repair and cell cycle control. *Protein Cell* **2014**, *5*, 750–760. [CrossRef]
24. Badaracco, M.E.; Ortiz, E.H.; Soto, E.F.; Connor, J.; Pasquini, J.M. Effect of transferrin on hypomyelination induced by iron deficiency. *J. Neurosci. Res.* **2008**, *86*, 2663–2673. [CrossRef]
25. Sprinzl, G.M.; Riechelmann, H. Current trends in treating hearing loss in elderly people: A review of the technology and treatment options—A mini-review. *Gerontology* **2010**, *56*, 351–358. [CrossRef]
26. Ivory, R.; Kane, R.; Diaz, R.C. Noise-induced hearing loss: A recreational noise perspective. *Curr. Opin. Otolaryngol. Head Neck Surg.* **2014**, *22*, 394–398. [CrossRef]
27. Tikka, C.; Verbeek, J.H.; Kateman, E.; Morata, T.C.; Dreschler, W.A.; Ferrite, S. Interventions to prevent occupational noise-induced hearing loss. *Cochrane Database Syst. Rev.* **2017**, *7*, CD006396. [CrossRef]
28. Breglio, A.M.; Rusheen, A.E.; Shide, E.D.; Fernandez, K.A.; Spielbauer, K.K.; McLachlin, K.M.; Hall, M.D.; Amable, L.; Cunningham, L.L. Cisplatin is retained in the cochlea indefinitely following chemotherapy. *Nat. Commun.* **2017**, *8*, 1654. [CrossRef]
29. Corbacella, E.; Lanzoni, I.; Ding, D.; Previati, M.; Salvi, R. Minocycline attenuates gentamicin induced hair cell loss in neonatal cochlear cultures. *Hear Res.* **2004**, *197*, 11–18. [CrossRef]

30. Khan, A.S.; Sheikh, Z.; Khan, S.; Dwivedi, R.; Benjamin, E. Viagra deafness–sensorineural hearing loss and phosphodiesterase-5 inhibitors. *Laryngoscope* **2011**, *121*, 1049–1054. [CrossRef]
31. Joo, Y.; Cruickshanks, K.J.; Klein, B.E.K.; Klein, R.; Hong, O.; Wallhagen, M. The contribution of ototoxic medications to hearing loss among older adults. *J. Gerontol. A Biol. Sci. Med. Sci.* **2019**, *75*, 561–566. [CrossRef]

International Journal of
Environmental Research and Public Health

Review

Manipulation of Lateral Pharyngeal Wall Muscles in Sleep Surgery: A Review of the Literature

Giovanni Cammaroto [1,2,*], Luigi Marco Stringa [3], Giannicola Iannella [1], Giuseppe Meccariello [1], Henry Zhang [4], Ahmed Yassin Bahgat [5], Christian Calvo-Henriquez [2,6], Carlos Chiesa-Estomba [2,7], Jerome R. Lechien [2,8], Maria Rosaria Barillari [2,9], Bruno Galletti [10], Francesco Galletti [10], Francesco Freni [10], Cosimo Galletti [11] and Claudio Vicini [1,12]

1 Head and Neck Department, ENT & Oral Surgery Unit, G.B. Morgagni, L. Pierantoni Hospital, Forlì, FC 47100 ASL of Romagna, Italy; giannicolaiannella@hotmail.it (G.I.); giuseppemec@yahoo.it (G.M.); claudio@claudiovicini.com (C.V.)
2 Young Otolaryngologists-International Federations of Oto-rhinolaryngological Societies (YO-IFOS), 75000 Paris, France; christian.calvo.henriquez@gmail.com (C.C.-H.); chiesaestomba86@gmail.com (C.C.-E.); Jerome.LECHIEN@umons.ac.be (J.R.L.); mariarosaria.barillari@unicampania.it (M.R.B.)
3 Department of Otolaryngology, Head and Neck Surgery, University of Ferrara, FE 44121 Ferrara, Italy; luigi.stringa@gmail.com
4 Department of Otolaryngology, Head and Neck, Royal London Hospital, London E1 1FR, UK; henryzhang87@gmail.com
5 Department of Otorhinolaryngology, Alexandria University, Alexandria 21526, Egypt; ahmedyassinbahgat@gmail.com
6 Department of otolaryngology, University Hospital Complex of Santiago de Compostela, 15706 Santiago de Compostela, Spain
7 Department of Otorhinolaryngology, Head & Neck Surgery, Hospital Universitario Donostia, 20014 San Sebastian, Spain
8 Department of Otolaryngology, Head & Neck Surgery, Foch Hospital, School of Medicine, UFR Simone Veil, Université Versailles Saint-Quentin-en-Yvelines (Paris Saclay University), 75000 Paris, France
9 Department of Mental and Physical Health and Preventive Medicine, University of L. Vanvitelli, CE 81100 Naples, Italy
10 Department of Adult and Development Age Human Pathology "Gaetano Barresi", Unit of Otorhinolaryngology, University of Messina, ME 98125 Messina, Italy; bgalletti@unime.it (B.G.); fgalletti@unime.it (F.G.); ffreni@unime.it (F.F.)
11 Comprehensive Dentistry Department, Faculty of Dentistry, Universitat de Barcelona, L'Hospitalet de Llobregat (Barcelona), 08907 Catalonia, Spain; cosimo_galletti@hotmail.it
12 ENT department, University of Ferrara, FE 44121 Ferrara, Italy
* Correspondence: giovanni.cammaroto@hotmail.com; Tel.: +39-054-363-5651

Received: 1 July 2020; Accepted: 22 July 2020; Published: 23 July 2020

Abstract: Background: Obstructive sleep apnea syndrome (OSAS) occurs due to upper airway obstruction resulting from anatomical and functional abnormalities. Upper airway collapsibility, particularly those involving the lateral pharyngeal wall (LPW), is known to be one of the main factors contributing to the pathogenesis of OSAS, leading the authors of the present study to propose different strategies in order to stiffen the pharyngeal walls to try to restore normal airflow. Methods: An exhaustive review of the English literature on lateral pharyngeal wall surgery for the treatment of OSAS was performed using the PubMed electronic database. Results: The research was performed in April 2020 and yielded approximately 2000 articles. However, considering the inclusion criteria, only 17 studies were included in the present study. Conclusions: The analyzed surgical techniques propose different parts of LPW on which to focus and a variable degree of invasivity. Despite the very promising results, no gold standard for the treatment of pharyngeal wall collapsibility has been proposed. However, thanks to progressive technological innovations and increasingly precise data analysis, the role of LPW surgery seems to be crucial in the treatment of OSAS patients.

Keywords: OSA; pharyngoplasty; sleep surgery; pharynx

1. Introduction

Obstructive sleep apnea syndrome (OSAS) is a very common health problem characterized by absent or insufficient ventilation during sleep as a consequence of the multilevel structural collapse of the upper airways, which usually involves the velopharynx, base of the tongue, and lateral pharyngeal walls. The diagnosis of OSAS is the result of the integration of anamnestic and clinical evaluations with instrumental data provided by polysomnography (PSG), which, collecting physiological parameters during sleep, allows a deeper and objective assessment of sleep-related breathing disorders. In order to localize the obstacle compromising correct airflow, a specific examination (like Müller's maneuver) can be performed when undergoing fiberoptic laryngoscopy and drug-induced sleep endoscopy (DISE). Described by different authors as a key point in the determination of airflow, the lateral pharyngeal wall (LPW) has been demonstrated to present an increased thickness and collapsibility in patients affected by OSAS, resulting in a potential cause of airway obstruction [1–4]. LPW includes muscular structures such as the palatoglossus muscle (PGM), the palatopharyngeal muscle (PPM), the superior pharyngeal constrictor (SPC), and, ultimately, lymphatic tissue, the palatine tonsils. Stabilizing LPW might be achieved by means of mandibular advancement devices or surgery [5]. Different authors have proposed surgical solutions to prevent the upper airways from collapsing [6], but in 2003, Cahali [7], for the first time, described lateral pharyngoplasty (LP), i.e., a surgical transoral approach focused on LPW, allowing, in this way, an enlarging of the oropharyngeal space and a stiffening of fibromuscular structures. Following on, surgeons have developed several techniques targeting the modification of LPW, trying to reduce surgical invasivity and complications, but none of them have been indicated as a gold standard procedure by the scientific community so far.

The efficacy of palate surgery is well documented by a meta-analysis published in 2018 by Pang et al. [8]. However, no reviews that are mainly oriented on lateral pharyngoplasty have been published so far. The aim of our study is to present a systematic review of the literature regarding the surgical techniques involving modifications of LPW for the treatment of OSAS in order to show their surgical stages, technical aspects, as well as their advantages and inconveniences, to better understand their applications on and contributions to the correction of upper airway obstruction.

2. Materials and Methods

A thoughtful review of the English language literature on the surgical procedures targeting LPW for the management of sleep apnea was performed using PubMed, EMBASE, Cochrane, and CENTRAL electronic databases. Two searches using (1) pharyngoplasty sleep apnea and (2) palate surgery sleep apnea as keyword clusters were performed, and they were combined with the use of the AND function to better select the research. Each paper included in the study met the following inclusion criteria: (1) the surgical treatment of OSAS patients, (2) the application of surgical procedures designed for structural modification of LPW, (3) the presence of instrumental parameters acquired by PSG during preoperative and postoperative evaluations, (4) reports including patients previously treated with Continuous Positive Airway Pressure (CPAP), and (5) reports including patients who did not undergo previous sleep surgery procedures. Articles that were not in accordance with the inclusion criteria were excluded. The subsequent criteria were applied with the aim of excluding inappropriate studies: surgical technique reports without significant outcome data, papers consisting of meta-analyses or literature reviews on surgical procedures, studies relative to ablative techniques, and articles describing animal or cadaveric samples. In order to further reduce the risk of incomplete literature analysis, a manual search through the bibliography of the included papers was carried out.

With the purpose of obtaining an organized and exhaustive review, the Preferred Reporting Items for Systematic Reviews and Meta-Analyses (PRISMA) criteria were applied first to select the papers and, secondly, to stratify them according to their level of evidence [9].

3. Results

In April 2020, the research led to the detection of approximately 2000 articles (from 1980 to 2020), but in accordance with the inclusion and exclusion criteria, 17 articles were enrolled in the present review. As the articles show detailed descriptions of surgical procedures to better clarify their singularities and effects, we grouped each article in accordance to which component of LPW is mostly involved and how deeply its modification occurs. As a result, we found four articles presenting surgical procedures in which SPC is manipulated and/or dissected, while in the remaining papers, PPM is the main structural target. In particular, five studies proposed a total or subtotal section of the muscle, four papers a partial section of the muscular fibers, and in four studies, minimal tissue handling is performed (Table 1, Figure 1).

Data on the efficacy and complications of LPW surgeries are shown in Table 1. The majority of samples included in the study consisted of patients affected by severe OSAS (mean pre-op apnea–hypopnea index (AHI) > 30). A mean reduction of more than 50% of AHI and a mean post-op AHI < 20 were observed in all series except one [10]. Some authors did not specify complications and their rates.

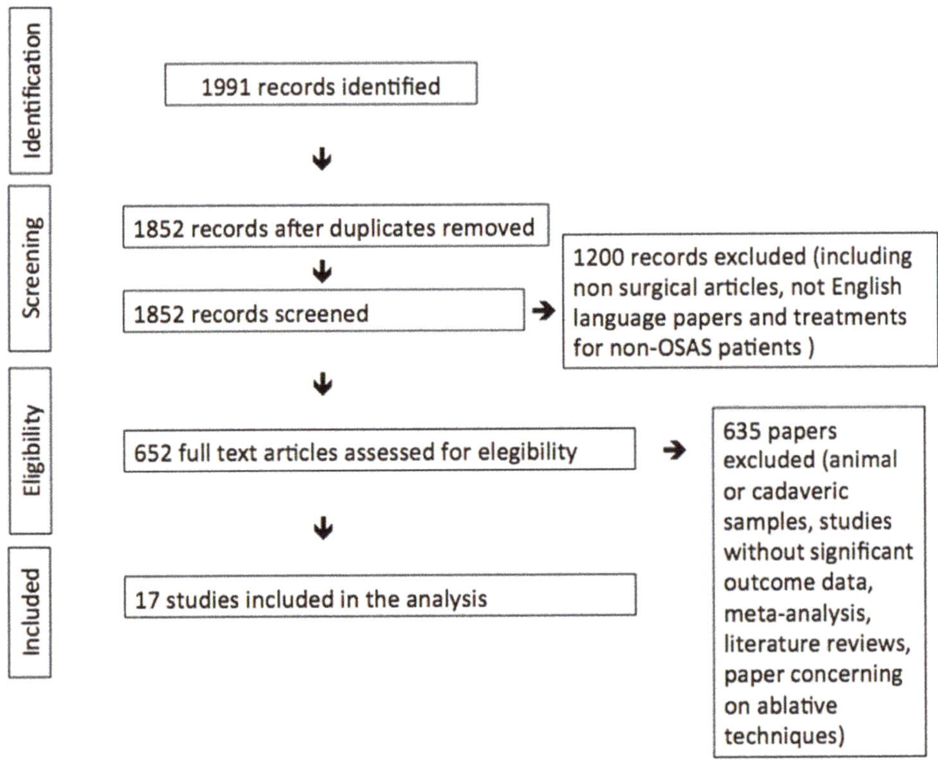

Figure 1. Preferred Reporting Items for Systematic Reviews and Meta-Analyses (PRISMA).

Table 1. Clinical data of papers included in the review.

Title	Structural Target and Modification of LPW	Number of Patients	Mean AHI Pre-Op	Mean AHI Post-Op	Δ AHI.	Follow-Up Time (Months)	Complications Described and Relating Incidence
Lateral Pharyngoplasty: A New Treatment for Obstructive Sleep Hypopnea Syndrome [7]	SPC	10	45.8	15.2	30.6	8.2	Oronasal reflux (10%), taste loss (10%).
Lateral-Expansion Pharyngoplasty: Combined Technique for the Treatment of Obstructive Sleep Apnea Syndrome [10]	SPC	38	22.4	13.6	8.8	7	-
Relocation Pharyngoplasty for Obstructive Sleep Apnea [11]	SPC	10	43.4	15.7	27.7	6	-
Anterolateral Advancement Pharyngoplasty: A New Technique for Treatment of Obstructive Sleep Apnea [12]	SPC	38	42.1	16.3	25.8	6	-
Expansion Sphincter Pharyngoplasty: A New Technique for the Treatment of Obstructive Sleep Apnea [13]	PPM section	45	44.2	12.0	32.2	6.5	-
Functional Expansion Pharyngoplasty in the Treatment of Obstructive Sleep Apnea [14]	PPM section	85	33.3	11.7	21.6	6	Postsurgical bleeding (2.3%).
Soft Palatal Webbing Flap Palatopharyngoplasty for Both Soft Palatal and Obstructive Sleep Apnea: A New Innovative Technique without Tonsillectomy [15]	PPM section	28	46.1	11	35.1	6	Excessive postnasal discharge (10.7%) temporary velopharyngeal insufficiency (7.1%), sensation of oral dryness (25%).
Modified Barbed Soft Palatal Posterior Pillar Webbing Flap Palatopharyngoplasty [16]	PPM section	21	47.7	12.3	35.4	6	Temporary velopharyngeal insufficiency (5%), excessive postnasal discharge (19%).
Modified Expansion Sphincter pharyngoplasty for Treatment of Children with Obstructive Sleep Apnea [17]	PPM partial section	25	60.5	2.0	58.5	-	Postoperative bleeding (4%).
A Modified Uvulopalatal Flap with Lateral Pharyngoplasty for Treatment in 92 Adults with Obstructive Sleep Apnoea Syndrome [18]	PPM partial section	92	39.1	7.9	21,2	6	Nasal regurgitation, bleeding, dysphagia, Foreign body sensation.
Barbed Reposition Pharyngoplasty (BRP) for OSAHS: A Feasibility, Safety, Efficacy and Teachability Pilot Study. "We are on the giant's shoulders" [19]	PPM partial section	10	43.65	13.57	30.08	6	Foreign body sensation, partial thread extrusion (20%).
Technical Update of Barbed Pharyngoplasty for Retropalatal Obstruction in Obstructive Sleep Apnoea [20]	PPM partial section	17	29.9	5.4	24.5	6	Foreign body sensation.
Barbed Roman Blinds Technique for Treatment of Obstructive Sleep Apnea: How We Do It? [21]	PPM minimal handling	32	36.9	13.7	23.2	12	-
Expansion Pharyngoplasty by New Simple Suspension Sutures without Tonsillectomy [22]	PPM minimal handling	24	28.6	8.9	19.7	9	-
Suspension Palatoplasty for Obstructive Sleep Apnea-Preliminary Study [23]	PPM minimal handling	25	39.8	15.1	24.7	6	Globus sensation of the throat.
Barbed Suspension Pharyngoplasty for Treatment of Lateral Pharyngeal Wall and Palatal Collapse in Patients Affected by OSAHS [24]	PPM minimal handling	20	25	5	20	6	Transient velopharyngeal insufficiency (10%), Partial thread extrusion (25%).

4. Discussion

As suggested by the literature, some OSAS patients present structural and functional abnormalities in the upper airways. In these patients, the pharynx (specifically, LPW) is thicker and more collapsible upon exposure to negative pressure during inspiration [4,5]. In 1981, Fujita et al. [6] introduced, for the first time, the uvulopalatopharyngoplasty (UPPP)—a surgical procedure finalized at the enlargement of the pharyngeal space—but the association of postoperative complications and dissatisfactory results drove surgeons to alternative approaches. Thus, LPW acquired a central role in OSAS pathogenesis, drawing the attention of numerous studies. Several surgical techniques have been proposed to enlarge and stiffen the pharyngeal tract. All of the procedures are performed under general anesthesia, with either nasotracheal or orotracheal intubation, while the patient is placed in the supine position and a mouth gag is applied to expose the oropharynx. What differentiates these techniques is the degree of involvement of the different LPW components, and how radical their modifications are.

Taking into account the sample sizes and the average postoperative follow up of the included articles, the following findings need to be highlighted.

Our review shows that the majority of the evaluated articles included patients with severe OSAS who experienced a significant postoperative reduction of AHI. These data show that the severity of sleep apneas should not be considered an exclusion criterion for LPW surgery.

However, the selection process of surgical candidates in each study was partially investigated in this review. The choice of performing preoperative DISE might influence the outcomes, and therefore, more attention should be dedicated to this aspect in future studies.

In our opinion, the main strength of our review is its specific focus on the muscular structures that are manipulated in each surgical technique. A better understanding of the involvement of the LPW muscles and prospective comparative studies might allow sleep surgeons to select the most effective surgical procedure.

4.1. Superior Pharyngeal Constrictor

In 2003, Cahali [7] presented LP as the first surgical procedure targeted to modify LPW for the treatment of pharyngeal instability, and SPC was the principal muscular structure involved. More precisely, after the individuation of SPC by bilateral tonsillectomy or mucosal removal of tonsillar fossa in the case of a previous tonsillectomy, it is split by a craniocaudal section into two flaps, one medially based, and the other, laterally based. The latter is sutured anteriorly to PGM. In order to obtain more space and simultaneously reduce traction forces, a palatal incision is performed from the lateral base of the uvula, extending laterally and superiorly, leading to the isolation of the upper part of PPM, which is partially sectioned in this portion. After that, two flaps are obtained: one superior flap, which is sutured in a Z-plasty fashion with the palatal flap, and one inferior flap, which is sutured to the anterior tonsillar pillar. The same procedure is performed on both sides. Although this innovative technique has shown satisfactory results in the correction of upper airway obstruction, it has been described to be related to transient oronasal reflux and partial taste loss. Despite the possible swallowing complication presented by LP, other authors consider SPC a successful target. José Antonio Pinto et al. [10], in so-called lateral-expansion pharyngoplasty, suggested the association of LP with expansion sphincter pharyngoplasty, presented for the first time by Pang and Woodson. In this way, the SPC section is combined with the section of PP muscle and the further suture of its cranial flap to the hamulus of the pterygoid process in order to improve the retropalatal obstruction. In their series of 38 samples, no patient suffered major complications. Regarding the modification of SPC, not all authors have pursued its section, and in some cases, a more conservative approach has been suggested. Huseh-Yu Li and Li-Ang Lee [11] introduced relocation pharyngoplasty, in which, once identified, SPC is grasped and sutured to PGM. At the same time, an elliptical cut from the lateral base of the uvula, extending superolaterally, is performed and the mucosal and submucosal adipose tissue is removed in order to gain further space, reducing tension forces. Once PPM is isolated from SPC, the posterior pillar is sewn to PGM. A distal mucosal resection of the uvula is also associated. Likewise, in anterolateral advancement pharyngoplasty by Emara et al., a relocation of SPC is performed without any muscular fiber section [12]. In particular, a limited dissection and partial separation of anterior and posterior parts of PPM from SPC were performed in the upper part of the tonsillar fossa. Then, SPC was plicated with a mattress-style suture and together with PPM (just inferiorly to the confluence point of its anterior and posterior parts) are anchored to the pterygomandibular raphe. Finally, the upper half of the posterior part of PPM is sutured to the levator veli palatini muscle.

4.2. Palatopharyngeal Muscle

After the very promising results showed by LP, many authors started to elaborate on new approaches aimed at the same outcomes, trying to avoid the complications. The manipulations of SPC were attributed as a significant source of postoperative swallowing problems, and in this way, PPM was considered a proper target in order to strengthen LPW. As shown below, a complete muscular fiber section is not performed in all the procedures, and a variable degree of tissue sparing is applied.

4.2.1. Total/Subtotal Section of Muscular Fibers

Three years after the introduction of LP, devising a surgical procedure that is easy to perform with a low rate of complications, K. Pang and T. Woodson [13] introduce the so-called expansion sphincter pharyngoplasty (ESP). In this new approach, all patients are submitted to bilateral tonsillectomy in order to identify PPM. Then, its muscular fibers are sectioned at the inferior end, and the posterior surface of the superior flap is partially detached from SPC. In order to isolate the soft palate muscles, a superolateral incision is performed on the anterior pillar. PPM is then lifted superolaterally, suturing it to the soft palate muscles. A partial uvulectomy is then performed. This technique showed very encouraging results and traced the path to more innovative approaches. A conservative modification of ESP was presented in 2013 by G. Sorrenti and O. Piccin, named functional expansion pharyngoplasty (FEP) [14]. The most important variation applied to the original technique is the replacement of the superolateral incision of soft palatal mucosa to expose the palatal muscles with a preparation of a tunnel through the palatal musculature from the apex to the hamulus of the tonsillar fossa of the pterygoid process. Once sectioned and rotated, the PPM flap is elevated through the tunnel and anchored to the palatine muscles close to the hamulus. A thin rim of muscular fibers of PPM is preserved medially to avoid retracting scar tissue at the posterior pillar. In this way, a more physiological widening force is applied to LPW, reducing tissue dissection and the subsequent retracting scar. Some years later, Sorrenti et al. [25] added some technical updates to their technique. In particular, they proposed the use of knotless barbed V-Loc sutures to fix the PPM flap to the apex of the mandibular pterygoid fold. Compared to previous FEP, this technique became easier and faster to perform with a powerful docking site. In order to reduce tissue manipulation and postoperative complications linked to tonsillectomy, A.M.M.E. Albassiouny, in 2014 [15], introduced the so-called soft palatal posterior pillar webbing flap palatopharyngoplasty technique in which the palatine tonsils are spared, and a transverse section of PPM is performed. Specifically, once the ventral mucosa of the posterior pillar is removed, it is sectioned in two points—one lateral, including PPM, and one medial, close to the uvula—in order to obtain two flaps. The lateral flap is then sutured to the most superolateral part of the anterior pillar, lateralizing the tonsil. After a submucosal dissection of the palatal mucosa and a shortening of the medial flap in length, it is turned up and sutured to the free margin of the soft palate. Two years later, A.M.M.E. Albassiouny [16] proposed two modifications to his original approach: coblation-assisted extracapsular tonsillectomy when tonsil collapse is documented by DISE, and the use of barbed STRATAFIX sutures in order to better control the tension impressed on the pharyngeal structure. In both cases, postoperative temporary velopharyngeal insufficiency and excessive postnasal discharge have been reported.

4.2.2. Partial Section of Muscular Fibers

The progressive interest in the treatment of OSAS has led to the elaboration of an increasingly conservative approach, allowing its application to a more delicate population such as pediatrics. On the heels of the excellent outcomes shown first by K. Pang and T. Woodson, and later by G. Sorrenti and O. Piccin, S.O. Ulualp [17] presented a modified ESP, applied to a pediatric population. In fact, in order to reduce tissue removal, after bilateral tonsillectomy, a partial section of the PPM anterior fibers is performed at the junction of the upper third and mid-third. Then, superolateral tunneling of the soft palate is prepared, and the upper portion of PPM is pulled up in the muscular tunnel and sutured to it. Not only the lateral side but also the anterior portion of the soft palate has been reported to be modified during LPW surgery. In this regard, M.J. Kim et al. [18] introduced a modified uvulopalatal flap with a partial LP in order to widen the retropalatal space anteroposteriorly and transversely. The posterior pillar is cut at its junction with the uvula, and once trimmed, it is sutured to the anterior pillar. After having calculated the mucosa and submucosal fat to be removed by measuring the halfway point between the junction of the soft and hard palates and the tip of the uvula, a diamond-shaped area of the soft palate is removed. Then, a suture of the muscular and mucosal rims is performed. In some cases, a bilateral tonsillectomy is performed, and the patients are submitted to concomitant nasal septoplasty and turbinate surgery. Following the interest in conservative approaches, and thanks to material technology improvements, in 2015, C. Vicini et al. [19] introduced an innovative tissue-sparing approach based on the use of barbed knotless bidirectional reabsorbable sutures named barbed reposition pharyngoplasty (BRP). All patients are submitted to bilateral tonsillectomy, and once PPM is demarcated, a partial incision is made at its inferior portion. A triangular-shaped mucosal and submucosal portion of the tonsillar fossa at its apex is removed in order to widen the oropharyngeal inlet. Then, the barbed sutured is utilized to superolaterally add tension to the upper portion of PPM and to stiffen LPW and the palate. In particular, the first pass of the needle is introduced at the center of the palate and passed laterally, reaching the pterygomandibular raphe. Passing around the pterygomandibular raphe, the needle reaches the upper portion of PPM through the tonsillar fossa. In this way, the muscular fibers are pulled up and anchored at the pterygomandibular raphe. The same procedure is applied to the other side. Effective and simple to perform, this technique has shown important results as both a single-stage procedure and a multilevel procedure associated with other upper airway corrective surgeries like base of tongue resection, hyoid suspension, and nasal surgery [26,27]. To minimize mucosal and muscular resection of the uvula, M.A. Babademez et al. [20] suggested some modifications using a monodirectional reabsorbable barbed thread to expand the upper part of the oropharyngeal inlet in a more conservative manner. In fact, in order to pull the uvula forward and superiorly, once PPM is fixed to the pterygomandibular raphe, the suture is passed horizontally through the root of the uvula to reach the opposite tonsillar bed, and then, PPM is anchored to the pterygomandibular raphe in the same fashion.

4.2.3. Minimal Handling of Muscular Fibers

In order to further reduce the complications linked to muscle fiber section and fibrotic tissue development, some authors have proposed some minimally invasive and nonresective procedures. In 2015, Mantovani et al., modifying their previous "Roman blinds technique" for the treatment of retropalatal collapse [21,28], presented the barbed Roman blinds technique (BRBT), in which complete preservation of oropharyngeal fibromuscular structures is encouraged, and stiffening of LPW and of the soft palate is obtained through the use of a bidirectional barbed suture. With the aim of exposing the muscular fibers of PPM, a mucosectomy of the tonsillar fossa is performed. The first needle of the bidirectional barbed suture is inserted in the palatal mucosa 1 cm in front of the posterior nasal spine. The suture is passed inferiorly, following the periosteal layer, to reach the periuvular extremity of PPM. Once the muscular fibers are grasped, the suture is fixed to the pterygoid hamulus and passed in a craniocaudal direction, encircling the pterygomandibular raphe. The needle is then passed medially, reaching PPM to anchor it to the pterygomandibular raphe. Once the muscular fibers are fixed laterally, the suture is directed to the hamulus, and then, to the first insertion point. A nonresective technique, barbed suspension pharyngoplasty (BSP), was also presented in 2019 by M. Barbieri et al. [24]. After a tonsillar fossa mucosectomy or a bilateral tonsillectomy, as in BRBT, no further tissue dissection is executed, but a bidirectional barbed suture is performed in order to lateralize LPW and to give more tension to the soft palate. In particular, the needle is first inserted in the palatal mucosa at the level of the posterior nasal spine and passed anterolaterally towards the upper part of the tonsillar fossa. Then, once PPM is anchored by multiple stitches to the anterior pillar, the suture is passed laterally towards the pterygomandibular raphe and then it medially encircles the contralateral raphe and is directed back to the ipsilateral raphe, passing through the base of the uvula. In this way, a stronger tension is applied to the soft palate, and a more secure suture is achieved. As previously reported, the pterygomandibular raphe is considered an important security point by many authors, and some of them perform a tissue dissection in order to expose it, so as to better distribute the strongpoints during the suture. Hsueh-Yu Li et al. [23] proposed a suspension palatoplasty, in which, after a mucosal incision from the anterior pillar rim to 1 cm in front of the center mark of the pterygomandibular raphe, the submucosal tissue is dissected to expose the fibers of the raphe and a bilateral tonsillectomy is performed. The upper third of PPM is then fixed at the pterygomandibular raphe at different points in a craniocaudal direction, and the anterior and posterior pillars are sutured together. No major complications have been reported except for a transient globus sensation. Not all the procedures acting to stabilize LPW have considered the removal of palatal lymphoid tissue as a required step. In this way, for those patients with LPW collapse without tonsillar hypertrophy, M.A. El-Ahl and M.W. El-Anwar [22] proposed a modified expansion pharyngoplasty without tonsillectomy and any pharyngeal tissue ablation, in which a reabsorbable submucosal suture is performed in order to fix the upper portions of PPM and PGM to the pterygoid hamulus.

In all the papers included in the review, a statistically significant improvement of the apnea–hypopnea index (AHI) was reported, showing a considerable impact of LPW surgery in the restoration of efficient airflow. Some complications have been reported, but none of them were of a major nature. Nasal regurgitation, dysphagia, foreign body sensation, velopharyngeal insufficiency, taste loss, and sensation of oral dryness have been the most noted postoperative symptoms, which resolved spontaneously after few weeks of the surgery. In a few cases, postoperative bleeding was reported, and, when a barbed suture was used, partial suture extrusion was described, without any additional problems. To better evaluate the surgical contributions to the amelioration of LPW collapse, some limits of the examined papers need to be mentioned. The small number of samples and the retrospective modality of the study are the most significant restrictions encountered, allowing, in this way, a partial estimation of the results. In addition, in some papers, the LPW surgical technique was presented as part of a multiple-level treatment and not as a single procedure, partially hiding the real impact on clinical improvement provided by the stiffening of the pharyngeal structures. Further efforts need to be made in order to clarify the authentic role of LPW surgery, and in this way, the quantification of upper airway collapsibility represents an important tool in order to perform an attentive selection of the candidate and thorough monitoring of the clinical data. The pharyngeal critical closing pressure (Pcrit) is the most accurate index of upper airway collapsibility, but the invasivity and the complexity of its measurement do not make it clinically practical. With regard to this, A.M. Osman et al. [29] recently presented the peak inspiratory flow percentage (PIF%) as a marker of the collapsibility of the upper airways; it is related to Pcrit and easy to obtain during a routine CPAP titration study. PIF% could represent an important opportunity to better quantify the structural collapsibility of the upper airways in clinical practice, allowing more direct estimation of the effective value of LPW surgery and optimizing, in this way, the treatment of OSAS patients.

Finally, the role of DISE as a phenotyping selection tool for surgical candidates also needs to be discussed. In fact, DISE appears to be a promising method for properly targeted therapy planning, in particular, allowing the selection or ruling-out of patients for specific surgical procedures [30].

5. Conclusions

Usually, upper airway collapse in patients with OSAS has a multilevel etiology, and LPW hypotonicity and flexibility represent two of the main factors. In this way, stiffening of the fibromuscular components and ablation of the redundant soft tissue of LPW constitute critical targets for many surgeons in order to restore correct airflow. Several surgical procedures acting on different fibromuscular structures have been proposed, each with different degrees of invasivity. With the improvements in surgical materials and knowledge acquisition around the physiopathology of LPW, surgeons have elaborated on increasingly efficient techniques to reduce the extension of tissue dissection and ablation. Many of them have reported important results, and while in some cases, postoperative complications have been experienced, these are usually temporary and of a small nature. Despite the high variability between the methods, all authors agree that meticulous preoperative analysis and selection of patients will reduce, as much as possible, the surgical failures. In this way, the absence of a gold standard, the possibility of utilizing the surgery of LPW in a multilevel context, the presence of a wide range of surgical options, and the short learning curve of many of them represent an opportunity to apply the most suitable method, according to the anatomical and clinical characteristics of the patient, to restore proper airflow.

Author Contributions: Conceptualization: G.C. and C.V.; data curation: G.I., G.M., and C.V.; formal analysis: L.M.S., C.C.-H., and C.C.-E.; investigation: G.C., G.I., and C.G.; methodology: G.I., F.F., and M.R.B.; project administration: G.C. and A.Y.B.; resources: G.C. and L.M.S.; supervision: H.Z. and J.R.L.; validation: C.V., F.G., and B.G. All authors have read and agreed to the published version of the manuscript.

Funding: This research received no external funding.

Conflicts of Interest: The authors report no potential conflict of interest.

References

1. Remmers, J.E.; DeGroot, W.J.; Sauerland, E.K.; Anch, A.M. Pathogenesis of upper airway occlusion during sleep. *J. Appl. Physiol.* **1978**, *44*, 931–938. [CrossRef] [PubMed]
2. Schwab, R.J.; Gefter, W.B.; A Hoffman, E.; Gupta, K.B.; Pack, A.I. Dynamic upper airway imaging during awake respiration in normal subjects and patients with sleep disordered breathing. *Am. Rev. Respir. Dis.* **1993**, *148*, 1385–1400. [CrossRef] [PubMed]
3. Isono, S.; Remmers, J.E.; Tanaka, A.; Sho, Y.; Nishino, T. Static properties of the passive pharynx in sleep Apnea. *Sleep* **1996**, *19*, 175–177.
4. Morrison, D.L.; Launois, S.H.; Isono, S.; Feroah, T.R.; Whitelaw, W.A.; Remmers, J.E. Pharyngeal narrowing and closing pressures in patients with obstructive sleep apnea. *Am. Rev. Respir. Dis.* **1993**, *148*, 606–611. [CrossRef]
5. Flores-Orozco, E.I.; Tiznado-Orozco, G.E.; Díaz-Peña, R.; Orozco, E.I.F.; Galletti, C.; Gazia, F.; Galletti, F. Effect of a mandibular advancement device on the upper airway in a patient with obstructive sleep apnea. *J. Craniofac. Surg.* **2020**, *31*, e32–e35. [CrossRef]
6. Fujita, S.; Cinway, W.; Zorik, F.; Roth, T. Surgical correction of ana-tomic abnormalities in obstructive sleep apnea syndrome: Uvulopalatopharyngoplasty. *Otolaryngol. Head Neck Surg.* **1981**, *89*, 923–934. [CrossRef]
7. Cahali, M.B. Lateral pharyngoplasty: A new treatment for OSAHS. *Laryngoscope* **2003**, *113*, 1961. [CrossRef]
8. Pang, K.P.; Plaza, G.; Baptista J, P.M.; Reina, C.O.; Chan, Y.H.; Pang, K.A.; Pang, E.B.; Wang, C.M.Z.; Rotenberg, B. Palate surgery for obstructive sleep apnea: A 17-year meta-analysis. *Eur. Arch. Oto-Rhino-Laryngol.* **2018**, *275*, 1697–1707. [CrossRef]

9. Urrutia, G.; Bonfill, X. Declaración PRISMA: Una propuesta para mejorar la publicación de revisiones sistemàticas y metaanálisis. *Med. Clin.* **2010**, *135*, 507–511. [CrossRef]
10. Pinto, J.A.; De Godoy, L.B.M.; Nunes, H.D.S.S.; Abdo, K.E.; Jahic, G.S.; Cavallini, A.F.; Freitas, G.S.; Ribeiro, D.K.; Duarte, C. Lateral-expansion pharyngoplasty: Combined technique for the treatment of obstructive sleep apnea syndrome. *Int. Arch. Otorhinolaryngol.* **2020**, *24*, e107–e111. [CrossRef]
11. Li, H.-Y.; Lee, L.-A. Relocation pharyngoplasty for obstructive sleep apnea. *Laryngoscope* **2009**, *119*, 2472–2477. [CrossRef] [PubMed]
12. Emara, T.A.; Hassan, M.H.; Mohamad, A.S.; Anany, A.M.; Ebrahem, A.E. Anterolateral advancement pharyngoplasty: A new technique for treatment of obstructive sleep apnea. *Otolaryngol. Head Neck Surg.* **2016**, *155*, 702–707. [CrossRef] [PubMed]
13. Pang, K.P.; Woodson, B.T. Expansion Sphincter Pharyngoplasty: A new technique for the treatment of obstructive sleep apnea. *Otolaryngol. Head Neck Surg.* **2007**, *137*, 110–114. [CrossRef] [PubMed]
14. Sorrenti, G.; Piccin, O. Functional expansion pharyngoplasty in the treatment of obstructive sleep apnea. *Laryngoscope* **2013**, *123*, 2905–2908. [CrossRef]
15. Elbassiouny, A.M.M.E. Soft palatal webbing flap palatopharyngoplasty for both soft palatal and oropharyngeal lateral wall collapse in the treatment of snoring and obstructive sleep apnea: A new innovative technique without tonsillectomy. *Sleep Breath.* **2014**, *19*, 481–487. [CrossRef] [PubMed]
16. Elbassiouny, A.M.M.E. Modified barbed soft palatal posterior pillar webbing flap palatopharyngoplasty. *Sleep Breath.* **2016**, *20*, 829–836. [CrossRef]
17. Ulualp, S.O. Modified expansion sphincter pharyngoplasty for treatment of children with obstructive sleep apnea. *JAMA Otolaryngol. Head Neck Surg.* **2014**, *140*, 817–822. [CrossRef]
18. Kim, M.; Kim, B.; Lee, D.; Choi, J.; Hwang, S.; Park, C.; Kim, S.; Cho, J.; Park, Y. A modified uvulopalatal flap with lateral pharyngoplasty for treatment in 92 adults with obstructive sleep apnoea syndrome. *Clin. Otolaryngol.* **2013**, *38*, 415–419. [CrossRef]
19. Vicini, C.; Hendawy, E.; Campanini, A.; Eesa, M.; Bahgat, A.; Alghamdi, S.; Meccariello, G.; DeVito, A.; Montevecchi, F.; Mantovani, M. Barbed reposition pharyngoplasty (BRP) for OSAHS: A feasibility, safety, efficacy and teachability pilot study. "We are on the giant's shoulders". *Eur. Arch. Oto-Rhino-Laryngol.* **2015**, *272*, 3065–3070. [CrossRef]
20. Babademez, M.A.; Gul, F.; Kale, H.; Sancak, M. Technical update of barbed pharyngoplasty for retropalatal obstruction in obstructive sleep apnoea. *J. Laryngol. Otol.* **2019**, *133*, 622–626. [CrossRef]
21. Mantovani, M.; Rinaldi, V.; Torretta, S.; Carioli, D.; Salamanca, F.; Pignataro, L. Barbed Roman blinds technique for the treatment of obstructive sleep apnea: How we do it? *Eur. Arch. Oto-Rhino-Laryngol.* **2015**, *273*, 517–523. [CrossRef] [PubMed]
22. El-Ahl, M.A.S.; El-Anwar, M.W. Expansion pharyngoplasty by new simple suspension sutures without tonsillectomy. *Otolaryngol. Head Neck Surg.* **2016**, *155*, 1065–1068. [CrossRef] [PubMed]
23. Li, H.-Y.; Lee, L.-A.; Kezirian, E.J.; Nakayama, M. Suspension palatoplasty for obstructive sleep apnea–a preliminary study. *Sci. Rep.* **2018**, *8*, 4224. [CrossRef] [PubMed]
24. Barbieri, M.; Missale, F.; Incandela, F.; Fragale, M.; Barbieri, A.; Roustan, V.; Canevari, F.R.; Peretti, G. Barbed suspension pharyngoplasty for treatment of lateral pharyngeal wall and palatal collapse in patients affected by OSAHS. *Eur. Arch. Oto-Rhino-Laryngol.* **2019**, *276*, 1829–1835. [CrossRef]
25. Sorrenti, G.; Pelligra, I.; Albertini, R.; Caccamo, G.; Piccin, O. Functional expansion pharyngoplasty: Technical update by unidirectional barbed sutures. *Clin. Otolaryngol.* **2018**, *43*, 1419–1421. [CrossRef]
26. Montevecchi, F.; Meccariello, G.; Firinu, E.; Arigliani, M.; De Benedetto, M.; Palumbo, A.; Bahgat, Y.; Bahgat, A.; Saldana, R.L.; Marzetti, A.; et al. Prospective multicentre study on barbed reposition pharyngoplasty standing alone or as a part of multilevel surgery for sleep apnoea. *Clin. Otolaryngol.* **2018**, *43*, 483–488. [CrossRef]
27. Vicini, C.; Meccariello, G.; Montevecchi, F.; De Vito, A.; Frassineti, S.; Gobbi, R.; Pelucchi, S.; Iannella, G.; Magliulo, G.; Cammaroto, G. Effectiveness of barbed repositioning pharyngoplasty for the treatment of obstructive sleep apnea (OSA): A prospective randomized trial. *Sleep Breath.* **2020**, *24*, 687–694. [CrossRef]
28. Mantovani, M.; Minetti, A.; Torretta, S.; Pincherle, A.; Tassone, G.; Pignataro, L. The velo-uvulo-pharyngeal lift or "roman blinds" technique for treatment of snoring: A preliminary report. *ACTA Otorhinolaryngol. Ital.* **2012**, *32*, 48–53.

29. Osman, A.M.; Tong, B.K.; Landry, S.A.; Edwards, B.A.; Joosten, S.A.; Hamilton, G.S.; Cori, J.M.; Jordan, A.S.; Stevens, D.; Grunstein, R.R.; et al. An assessment of a simple clinical technique to estimate pharyngeal collapsibility in people with obstructive sleep apnea. *Sleep* **2020**. [CrossRef]
30. De Vito, A.; Agnoletti, V.; Zani, G.; Corso, R.M.; D'Agostino, G.; Firinu, E.; Marchi, C.; Hsu, Y.-S.; Maitan, S.; Vicini, C. The importance of drug-induced sedation endoscopy (D.I.S.E.) techniques in surgical decision making: Conventional versus target controlled infusion techniques—A prospective randomized controlled study and a retrospective surgical outcomes analysis. *Eur. Arch. Oto-Rhino-Laryngol.* **2017**, *274*, 2307–2317. [CrossRef]

International Journal of
Environmental Research and Public Health

Review

Pneumoparotid and Pneumoparotitis: A Literary Review

Francesco Gazia [1,*], Francesco Freni [1], Cosimo Galletti [1], Bruno Galletti [1], Rocco Bruno [1], Cosimo Galletti [2], Alessandro Meduri [3] and Francesco Galletti [1]

1 Department of Adult and Development Age Human Pathology "Gaetano Barresi", Unit of Otorhinolaryngology, University of Messina, 98125 Messina, Italy; franco.freni@tiscali.it (F.F.); cosimogalletti92@gmail.com (C.G.); bgalletti@unime.it (B.G.); brunor@unime.it (R.B.); fgalletti@unime.it (F.G.)

2 Comprehensive Dentistry Department, Faculty of Dentistry, Universitat de Barcelona, L'Hospitalet de Llobregat (Barcelona), 08907 Catalonia, Spain; cosimo88a@gmail.com

3 Department of Scienze Biomediche, Odontoiatriche e Delle Immagini Morfologiche e Funzionali, Unit of Ophthalmology, University of Messina, 98125 Messina, Italy; ameduri@unime.it

* Correspondence: ssgazia@gmail.com; Tel.: +39-0902212248; Fax: +39-0902212242

Received: 27 April 2020; Accepted: 26 May 2020; Published: 2 June 2020

Abstract: Pneumoparotid is a rare condition of parotid swelling. The presence of the air in gland parenchyma is caused by an incompetent Stensen's duct with high pressure may cause the acini's rupture. We reviewed 49 manuscripts, from 1987 to today, that enrolled a total of 54 patients with pneumoparotid. Our review evaluated the following evaluation parameters: gender, age, etiology, clinical presentation, treatment, days of resolution after diagnosis, relapse and complications. The most frequent etiology is self-induction by swelling the cheeks (53.7%). This cause mainly involves children (74%), for conflicts with parents, excuses for not going to school, nervous tics or adults (16%) with psychiatric disorders. Iatrogenic causes are also frequent (16.6%), for dental treatments (55.5%) or use of continuous positive airway pressure (CPAP) (33.4%). Medical therapy is the most practiced (53.7%), in most cases it is combined with behavioral therapy (25.9%) or psychotherapy (25.9%). Surgery is rarely used (9.2%) as a definitive solution through parotidectomy (50%) or ligation of the duct (50%). The most common complication is subcutaneous emphysema (24.1%), sometimes associated with pneumomediastinum (5.5%). Careful treatment and management are necessary to ensure the resolution of the pathology and counteract the onset of complications.

Keywords: pneumoparotid; pneumoparotitis; parotitis; Stensen's duct; head and neck

---

1. Introduction

Pneumoparotid is a rare cause of parotid enlargement due the presence of air within the parotid gland. The pneumoparotid term, first described in 1865 by Hyrtl, defines the presence of air within parotid system: gland and Stensen's duct [1]. The condition was recognized also in 1915 when a strange epidemic of mumps occurred in the French Foreign Legion in North Africa. The soldiers were deliberately self-inducing the condition by blowing into a small bottle to avoid duty [2]. Conditions that increase intraoral pressure like Valsalva's maneuver or incompetent Stensen's duct are predisposing factor to pneumoparotid. Pneumoparotitis is a complication of pneumoparotid that proceeds towards an inflammatory state or infection process. In general, local pain in the parotid area and swelling are the most common symptoms. We have noticed how often in the literature pneumoparotid and pneumoparotitis are used interchangeably. In reality, the latter is a complication of the former. In our review, we clarified the real percentage of this complication. Subcutaneous emphysema has been described as a complication of this condition and occurs from an extension of the air leak from the affected parotid acini to the surrounding cervicofacial subcutaneous tissues [3]. Literature shows cases of pneumoparotid in adolescents and adults with psychosocial issues. A correct anamnesis and imaging studies like ultrasound, sialendoscopy and head–neck computed tomography (CT) are essential to perform a correct diagnosis (Figure 1). Treatment generally includes supportive medical management, reserving surgical therapy in case of severe cases [4].

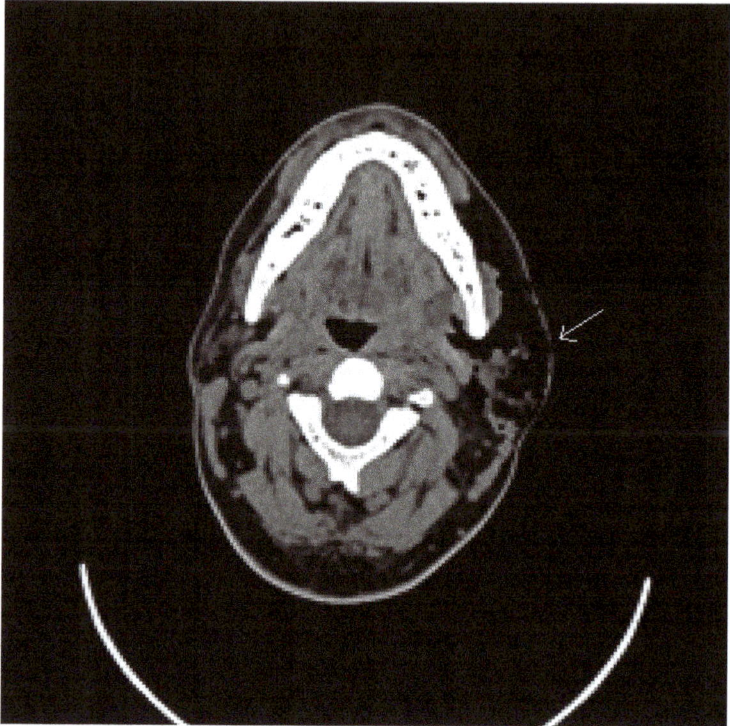

Figure 1. Axial projection computed tomography (CT) image of a left pneumoparotid case, with the arrow indicating the presence of air in the parotid lodge.

The main problem of pneumoparotid is that it is the clinical condition not well-described in the literature—only clinical reports are published, without any observational study with large numbers of patients, no studies comparing the various treatments or how to prevent complications.

The purpose of our review is to collect all the data present in the literature and make a general analysis on the epidemiology, etiology, treatment and management of this rare disease.

Furthermore, in the literature there are only case reports, we wanted to write the first review to clarify all the salient points of this clinical condition and to provide the scientific community with the correct indications to diagnose and quickly treat pneumoparotid, avoiding complications.

2. Materials and Methods

We have analyzed the case reports or case series in English, full-text access (open access or payment) that have pneumoparotid treatment and management as their main topic. All articles were found on PubMed, Scopus and Web of Science using the keywords "pneumoparotid", "pneumoparotidis", "pneumoparotis" and "parotid emphysema" in four different searches. The data of this systematic investigation observed the preferred reporting items for systematic review (PRISMA) accordingly with the statement (Figure 2). We only considered the cases of symptomatic patients, excluding patients with occasional findings (for example after the puffed-cheek maneuver for the CT study of the oral cavity). We reviewed 49 manuscripts, from 1987 to today, that enrolled a total of 54 patients with pneumoparotid. Our review evaluated the following evaluation parameters: gender, age, etiology, clinical presentation, treatment, days of resolution after diagnosis, relapse and complications (Table 1).

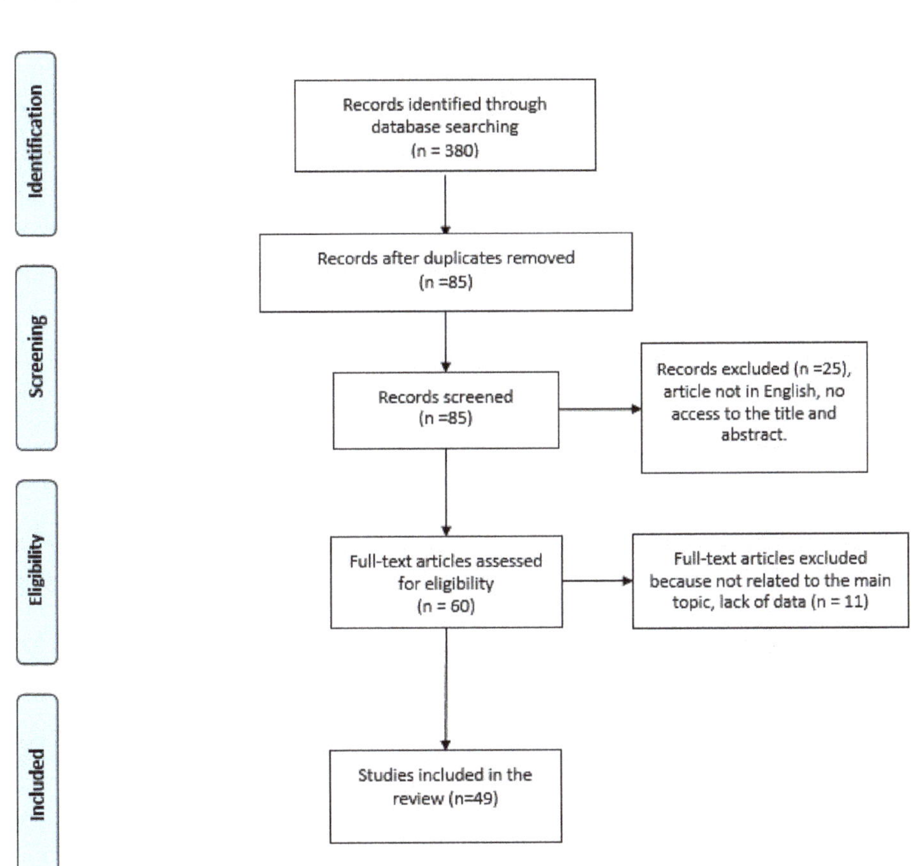

Figure 2. Review preferred reporting items for systematic review (PRISMA) flow diagram.

Table 1. Manuscripts analyzed.

Manuscript	Sex	Age	Clinical Presentation	Etiology	Treatment	Resolution after Diagnosis	Relapse	Complication
Garber et al., 1987 [5]	M	32	Bilateral	Hay fever (Coughing attack)	Medical	5 days	No	
Markowitz et al., 1987 [6]	F	12	Bilateral	Self-induced	Medical and psychotherapy	1 day	Yes	
David et al., 1988 [7]	F	6	Left	Self-induced	Medical, needle aspiration and psychotherapy		No	Parotitis
Brodie et al., 1988 [8]	M	14	Bilateral	Self-induced	Surgery (transposition of the duct)		No	Subcutaneous emphysema
Telfer et al., 1989 [9]	M	29	Right	Idiopathic	Surgery (treatment of drooling second Brody)		No	/
Mandel et al., 1991 [10]	M	53	Right	Self-Induced	Behavioral	1 day	Yes	/
Piette et al., 1991 [11]	F	34	Right	Iatrogenic (dental care)	Medical	5 days	No	/
Takenoshita et al., 1991 [12]	M	24	Left	Iatrogenic (dental care)	Medical	2 days	No	/
Krief et al., 1992 [13]	M	10	Bilateral	Self-induced	Medical and psychotherapy	/	Yes	Parotitis
Curtin et al., 1992 [14]	M	26	Bilateral	Self-Induced	Behavioral	/	No	/
Ferlito et al., 1992 [15]	M	14	Bilateral	Self-Induced	Medical and psychotherapy	/	Yes	
Brown et al., 1993 [16]	M	30	Left	Iatrogenic (Air-powder prophylaxis units for removing plaque)	Medical	5 days	No	/
Birzgalis et al., 1993 [17]	M	16	Right	Self-Induced	Behavioral	/	No	Subcutaneous emphysema

Table 1. *Cont.*

Manuscript	Sex	Age	Clinical Presentation	Etiology	Treatment	Resolution after Diagnosis	Relapse	Complication
McDuffie et al., 1993 [18]	M	24	Bilateral	Iatrogenic Orthodontic appliances	Behavioral (correction of orthodontic appliances)	3 days	No	/
Cook et al., 1993 [19]	F	44	Bilateral	Coughing attack	None		No	
Nassimbeni et al., 1995 [20]	M	12	Bilateral	Self-induced	Psychotherapy		No	Subcutaneous emphysema
	M	9	Right	Self-induced	Surgery (parotidectomy)		Yes	Abscess
Goguen et al., 1995 [21]	M	9	Right	Self-induced	Behavioral	1 day	No	
	F	9	Bilateral	Self-Induced	Medical and psychotherapy		Yes	Parotitis
	M	13	Bilateral	Self-Induced	Psychotherapy		Yes	
Ros et al., 1996 [22]	M	3	Left	Self-Induced	None	1 day	No	
Gudlaugsson et al., 1998 [23]	F	16	Bilateral	Self-induced	Medical and psychotherapy		Yes	Subcutaneous emphysema, pneumomediastinum
Alcalde et al., 1998 [24]	M	29	Right	Idiopathic	Needle aspiration, medical and bite		No	
Golz et al., 1999 [25]	M	10	Bilateral	Self-induced	Psychotherapy		No	
Sittel et al., 1999 [26]	F	14	Bilateral	Self-induced	Medical and psychotherapy		Yes	
Kirsch et al., 1999 [27]	M	41	Left	Iatrogenic Spirometry	None	1 day	Yes	
Martín-Granizo et al., 1999 [28]	F	5	Bilateral	Idiopathic	Medical	2 days	Yes	
	F	8	Right	Coughing attack	Medical			
Han et al., 2004 [29]	M	13	Rigt	Self-induced	Medical and surgery (duct ligation)	2 days	No	Subcutaneous emphysema

Table 1. *Cont.*

Manuscript	Sex	Age	Clinical Presentation	Etiology	Treatment	Resolution after Diagnosis	Relapse	Complication
Apaydin et al., 2004 [30]	M	50	Left	Idiopathic	Surgery (parotidectomy)	After surgery	No	
Grainger et al., 2006 [31]	F	12	Bilateral	Idiopathic	Medical	2 days	Yes	
Balasubramanian et al., 2008 [3]	M	11	Bilateral	Self-induced	Medical and psychotherapy		Yes	Subcutaneous emphysema
Prabhu et al., 2008 [32]	M	12	Bilateral	Self-induced	Medical and psychotherapy		Yes	
Luaces et al., 2008 [33]	M	11	Right	Self-induced	Medical	28 days	Yes	Subcutaneous emphysema
Faure et al., 2009 [34]	M	9	Left	Self-induced	Medical and psychotherapy		Yes	
Kyung et al., 2010 [35]	F	7	Bilateral	Self-induced	Medical and behavioral	3 days	No	Subcutaneous emphysema, pneumomediastinum
Zuchi et al., 2011 [1]	F	50	left	Idiopathic	Medical	14 days	No	Parotitis
Mukundan et al., 2011 [36]	M	13	Left	Self-induced	Medical		No	
van Ardenne et al., 2011 [37]	F	7	Right	Self-induced	Medical and Behavioral	30 days	No	
Ghanem et al., 2012 [38]	M	46	Unilateral	Idiopathic	Sialendoscopy		Yes	Parotitis
Potet et al., 2012 [39]	F	44	Left	Idiopathic	Medical		No	Parotitis
McGreevy et al., 2013 [2]	M	48	Right	Idiopathic	Surgery (parotidectomy)	After surgery		Parotitis (before surgery)
McCormick et al., 2013 [4]	M	7	Bilateral	Idiopathic	Medical		Yes	
Konstantinidis et al., 2014 [40]	M	61	Right	Idiopathic	Sialendoscopy with corticosteroids		Yes	Parotitis

Table 1. *Cont.*

Manuscript	Sex	Age	Clinical Presentation	Etiology	Treatment	Resolution after Diagnosis	Relapse	Complication
Abdullayev et al., 2014 [41]	M	36	Bilateral	Iatrogenic CPAP	Behavioral (stopping CPAP)	1 day	No	
Cabello et al., 2015 [42]	M	42	Right	Iatrogenic MAD	Behavioral (regulating MAD)		No	
Alnæs et al., 2017 [43]	F	10	Left	Self-induced	Medical and behavioral	1 day	Yes	Subcutaneous emphysema
Goates et al., 2017 [44]	M	53	Left	Iatrogenic CPAP	Behavioral (nasal CPAP)	1 day	No	
	M	54	Right	Iatrogenic CPAP	Behavioral (nasal CPAP)	1 day	No	
Lagunas et al., 2017 [45]	M	13	Bilateral	Self-induced	Medical and behavioral	1 day	Yes	Subcutaneous emphysema
Yamazaki et al., 2017 [46]	M	53	Bilateral	Self-induced	Medical and behavioral		No	
Lee et al., 2018 [47]	M	11	Bilateral	Idiopathic	Medical	4 days		Subcutaneous emphysema, pneumomediastinum
House et al., 2018 [48]	M	34	Bilateral	Self-induced	Medical and psychotherapy		Yes	Subcutaneous emphysema, parotitis
Ambrosino et al., 2019 [49]	M	12	Bilateral	Idiopathic	Medical		Yes	Subcutaneous emphysema

3. Results

We analyzed a total of 54 patients, 39 males and 15 females. The mean age was 22.3 years, but we can consider a group of patients in scholar age (31 patients, 11.9 years mean age) and a group of adults (23 patients, 40.8 years mean age). The clinical presentation is characterized by swelling of the parotid region sometimes extended to others districts, which can be bilateral (48.1%) or unilateral (51.9%). The most frequent etiology is self-induction by swelling the cheeks (53.7%). This cause mainly involves children (84%), for conflicts with parents, excuses for not going to school, nervous tics or adults (16%) with psychiatric disorders. The cases of idiopathic pathology are 24.1%. Iatrogenic causes are also frequent (16.6%), for dental factors (55.5%), use of CPAP (33.4%) or during spirometry (11.1%). Persistent coughing attacks can also be a cause in subjects with chronic bronchitis (5.5%).

Regarding the treatment, medical therapy is the most practiced (53.7%), with the use of antibiotics and steroidal anti-inflammatories or not. In most cases, medical therapy is combined with other treatments. Behavioral therapy is used to remove bad habits that can lead to this pathology (25.9%), with zeroing the recurrence rate if the subject is collaborative. If the subject has mental disorders, supportive psychotherapy is often required (25.9%), with a prevalence in children (95%). When the pathology does not resolve or tends to be recidive, more invasive approaches are used, such as needle aspiration (3.7%). Surgery is rarely used (11.1%) as a definitive solution through parotidectomy (50%) or ligation of the duct (50%). Corticosteroid infiltration sialoendoscopy was used in 2 cases without success. There were also 3 cases (5.5%) that did not require any treatment for resolution. Regarding our analysis, the pathology resolves in 4.5 days with the appropriate treatment, due to the low number of cases further investigation occurred. The disease relapsed in 23 subjects, but in 3/51 cases no data concerning the recurrence rate was found. From the data we analyzed, the recurrence rate is 42.6%, mainly affecting psychiatric subjects (60%). The most common complication is subcutaneous emphysema (24.1%), sometimes associated with pneumomediastinum (5.5%). Parotitis associated with pneumoparotide, which is called pneumoparotitis, has only been described in 14.8% of cases, underlining an improper use of this term. Abscess of the parotid lodge occurred only once (1.8%) (Table 2).

Table 2. Summary of results.

Results	**M ± SD** ***n* (%)**
Gender	
Male	39/54 (72.2%)
Female	15/54 (27.8%)
Age (Years)	22.3 ± 17.7
Clinical Presentation	
Bilateral	26/54 (48.1%)
Monolateral	28/54 (51.9%)
Etiology	
Self-induced	29/54 (53.7%)
Idiopathic	13/54 (24.1%)
Iatrogenic	9/54 (16.6%)
Coughing attack	3/54 (5.5%)
Treatment	
Medical	29/54 (53.7%)
Psychotherapy	14/54 (25.9%)
Behavioral	14/54 (25.9%)
Surgery	6/54 (11.1%)

Table 2. *Cont.*

Results	M ± SD *n* (%)
Needle aspiration	2/54 (3.7%)
Sialendoscopy	2/54 (3.7%)
None	3/54 (5.5%)
Resolution after Diagnosis (Days)	4.5 ± 7.8
Relapse	
Yes	23/54 (42.6%)
No	28/54 (51.8%)
Unspecified	3/54 (5.5%)
Complications	
Subcutaneous emphysema	13/54 (24.1%)
Pneumomediastinum	3/54 (5.5%)
Abscess	1/54 (1.8%)
Parotitis	8/54 (14.8%)

n, number;%, percentage; M, media; SD, standard deviation.

4. Discussion

Pneumoparotid is a very rare condition of parotid gland, often complicating with a subcutaneous emphysema, causing swelling of the parotid lodge. This pathology usually occurred due no physiological stagnation of air in parotid parenchyma. Pneumoparotid is usually associated with a retrograde insufflation of air and saliva via Stensen's duct into the secondary ducts and glandular acini [40]. Hypotonia of the buccinator muscle, hypertrophy of the masseter muscle or temporary obstruction of the Stensen's duct by mucous are described as possible risk factors [1].

The opening of the Stensen's duct lies near to the second upper molar tooth bilaterally. The normal anatomy of duct preventing the reflux of air and saliva into the parotid gland are three fold:

1. The diameter of the duct orifice is smaller than that of the duct itself;
2. The duct opening is covered by redundant mucosal layer, covering the duct orifice when there is increased intraoral pressure;
3. The Stensen's duct is compressed in its lateral course along the masseter muscle and its passage through the buccinator muscle with an increase in oral pressure.

In our experience, we report a case of a 45-year-old man with numerous episodes of painful, mono lateral left facial swelling. Clinical examination reported left-sided painful and parotid swelling with crepitus. Head–neck CT examination reported very important presence of subcutaneous emphysema that affected caudo-cranial left soft tissues from temporal region to the upper thoracic outlet, severe ectasia of Stensen's duct, ducts of salivary glands and left parotid (Figure 1). Aware of the patient's psychiatric conditions, psychiatric counseling is demanding. The colleagues reported that the patient suffered form of a minor cognitive disability with a tendency to somatization, underlying an important state of anxious and insomnia, prescribing a psychiatric therapy with venlafaxine, quetiapine and alprazolam. The patient is treated with antibiotic therapy and support measures with resolution of subcutaneous emphysema and general health condition. Our experience is in agreement with the case studies, management and treatment of the pathology described in the literature.

Medical literature showed a frequent association with glass blowing, playing wind instruments, exercising and self-induced behaviors often linked to psychiatric disorders. Normal intraoral pressure is 2 to 3 mm Hg, in glassblowing and trumpet playing this pressure may increase until 150 mmHg facilitating the disease's development. Furthermore, iatrogenic pneumoparotid is described like complication of spirometry, odontoiatric procedures, fine needle aspiration of the parotid gland and positive pressure ventilation used preoperatively or in the intensive care setting [11,16,18,27,

42,50–52]. Long-term use of oronasal continuous positive airway pressure is a potential cause of pneumoparotid [41,44]. Viral and bacterial infections, autoimmune diseases like sarcoidosis, Sjögren syndrome and Wegner's vasculitis, diabetes, Cushing disease, hypothyroidism, liver disease are described like possible causes of pneumoparotid or pneumoparotidis [2].

Repeated episodes of pneumoparotid may cause to chronic inflammation, infection or sialectasis.

The pathophysiologic condition of pneumoparotid has also been demonstrated by using a "puffed-cheek" technique [53], usually performed a CT examination after sialography, which mark filling defects, air in the parotid ductal system and sialoliths. Next, massaging the both patient's parotid glands, CT scan is performed highlighted a reduced amount of air and absence of contrast. Repeated maneuvers of autoinflation with high pressure may cause the acini's rupture. As we know the parotid's capsule is incomplete in the superiomedially part at the posterior border of mandible bone, airflow could reach the parapharyngeal and retropharyngeal space [2], provoking emphysema.

Enlargement of the parotid gland may be due to mumps, bacterial sialadenitis, obstructive sialadenitis, autoimmune disease like Sjogren syndrome. There are also rarer causes that can lead to swelling of the parotid, for example tuberculosis, sarcoidosis, cat-scratch disease or trauma. Pneumoparotid refers to the pathologic state of air within the parotid gland with or without inflammation. The clinical history of the patient (glass blowing, playing wind instruments, self-induced behaviors often linked to psychiatric disorders) and radiodiagnostics play a crucial role in the differential diagnosis. Pneumoparotid should be suspected with painless or minimally painful parotid swelling in the absence of fever. In the acute phase, plain radiographs may show air within the ductal system, sometimes with extravasation into the parenchyma and surround soft tissues. Computed tomography demonstrates air contrast with great sensitivity. Ductal dilation is a common finding on both sialography and computed tomography.

Imaging techniques are essential to perform a correct diagnosis. In reviewing the medical literature, radiologic studies that are indicated as good practice are ultrasonography, sialography, radionuclide sialography, sialendoscopy, salivary gland isotope scanning, CT and nuclear magnetic resonance (NMR) [54]. The use of ultrasound is stronglyrecommended in the diagnosis of superficial swelling in the head–neck area in general, and for salivary gland diseases in particular. It marks multiple hyperechoic areas corresponding to air in the glandular parenchyma, ducts and soft tissue. It is easy, reliable, non-invasive, cost-efficient and provides real-time conservative dynamic imaging. Scialography is useful for establishing the presence of stones, although less sensitive. [2,4].

In recent last years, sialendoscopy has become a good routine technique and minimally invasive diagnostic procedure of the parotid gland. The main goal is the evaluation and management of the salivary ductal system [40]. Currently, CT is the gold-standard technique because it defines anatomy and it is not invasive.Describing air-filled dilatation of Stensen's duct, glandular acini air dilatation, collections, free air intraparenchymal and a good imaging of duct glandular system, also helps in diagnoses of extension of air-accumulation in the nearest areas of the head–neck district [2]. Puffed-cheek CT is a good technique that demonstrated a subtle, but definite increase in intraductual and intraglandular parotid air when is compared to the simple CT [53].

Clinical treatment is the first step in approaching pneumoparotid. Acute management includes a short line of antibiotics, oral or intravenous, with the addition of steroids if the swelling is severe.

Antibiotics are used to protect the host from secondary infections; analgesia is also considered to improve general health state of patient. A parallel line of treatment includes massage of the gland, hydration, mouthwashes, sialogogues and warm compresses. In self-induced pneumoparotid cases, psychotherapy is necessary to correct the underlying adaptative psychiatric disorder. In severe cases or recurrences—sometimes associated with infection or pneumomediastinum—surgery is required: glandular resection, ductoplastic and/or Stensen's duct ligation, partial parotidectomy with duct's ligation. Parotid duct ligation is considered as a gold-standard for recurrent or chronic severe parotid infection. Parotidectomy is required in rare cases, usually when the patient is noncompliant, in failure of treatment or chronic infection, is the end point line of treatment [2,9,20,29,30,55].

Parotidectomy is an invasive surgery procedure that can induced complications that patients and professionals have to considered: partial or complete facial nerve lesion, Frey's syndrome [56,57], salivary fistula, auricularis magnus nerve lesion and keloid cicatrization of surgery incision. To avoid the recurrences of pneumoparotid a counseling to explain that are it is essential to stop activities that increase intraoral pressure is already fundamental.

The limit of our review is represented by the fact that all the selected articles are case reports or case series. There are no observational, retrospective or prospective studies in the literature. This review may be a starting point for clinical studies with a larger number of patients. Given the lack of comparative studies between the various therapeutic treatments or on the prevention of complications, further studies are needed for the definition of guidelines or gold-standard.

5. Conclusions

Pneumoparotid is not a real pathology, but a non-physiological clinical condition characterized by the presence of air in the Stensen's duct and throughout the gland—and can be complicated. Pneumoparotid affects two target populations, children and adults. Thanks to this review, we have clarified some important aspects concerning the etiopathogenesis and pathophysiology of pneumoparotid. We pointed out that the most frequent cause is self-induction, caused most often by people with psychiatric disorders. Regarding the treatment, there is no gold-standard, but each patient must be treated according to his/her clinical condition, speeding up the diagnostic process through a CT examination. In case of complications such as pneumoparotitis, antibiotic therapy is indispensable. In the complication of subcutaneous emphysema, the clinic, the size and the recurrence rate must always be evaluated to avoid the evolution towards pneumomediastinum. In case of critical dimensions, needle aspiration or surgical treatment is appropriate. In case of recurrence, more aggressive surgical treatment should be considered. Careful treatment and management are necessary to ensure the resolution of the pathology and counteract the onset of complications.

Author Contributions: Conceptualization, F.G. (Francesco Gazia) and C.G.1; methodology, F.G. (Francesco Gazia); software, R.B.; validation, B.G., F.F. and F.G. (Francesco Galletti); formal analysis, F.G. (Francesco Gazia); investigation, C.G.1 and A.M.; resources, F.G. (Francesco Gazia) and C.G.1; data curation, C.G.2; writing—original draft preparation, C.G.1 and F.F.; writing—review and editing, R.B. and B.G.; visualization, C.G.2; supervision, F.G. (Francesco Galletti); project administration, F.F., C.G.2 and A.M.; All authors have read and agreed to the published version of the manuscript.

Funding: This research received no external funding.

Conflicts of Interest: The authors declare no conflict of interest.

References

1. Zuchi, D.F.; Da Silveira, P.C.; Cardoso Cde, O.; Almeida, W.M.; Feldman, C.J. Pneumoparotitis. *Braz. J. Otorhinolaryngol.* **2011**, *77*, 806. [CrossRef] [PubMed]
2. McGreevy, A.E.; O'Kane, A.M.; McCaul, D.; Basha, S.I. Pneumoparotitis: A case report. *Head Neck* **2012**, *35*, E55–E59. [CrossRef] [PubMed]
3. Balasubramanian, S.; Srinivas, S.; Aparna, K.R. Pneumoparotitis with subcutaneous emphysema. *Indian Pediatr.* **2008**, *45*, 58–60. [PubMed]
4. McCormick, M.E.; Bawa, G.; Shah, R.K. Idiopathic recurrent pneumoparotitis. *Am. J. Otolaryngol.* **2013**, *34*, 180–182. [CrossRef]
5. Garber, M.W. Pneumoparotitis: An unusual manifestation of hay fever. *Am. J. Emerg. Med.* **1987**, *5*, 40–41. [CrossRef]
6. Markowitz-Spence, L.; Brodsky, L.; Seidell, G.; Stanievich, J.F. Self-induced pneumoparotitis in an adolescent. Report of a case and review of the literature. *Int. J. Pediatr. Otorhinolaryngol.* **1987**, *14*, 113–121. [CrossRef]
7. David, M.L.; Kanga, J.F. Pneumoparotid. In cystic fibrosis. *Clin. Pediatr. (Phila)* **1988**, *27*, 506–508.
8. Brodie, H.A.; Chole, R.A. Recurrent Pneumosialadenitis: A Case Presentation and New Surgical Intervention. *Otolaryngol. Head Neck Surg.* **1988**, *98*, 350–353. [CrossRef]
9. Telfer, M.R.; Irvine, G.H. Pneumoparotitis. *Br. J. Surg.* **1989**, *76*, 978.

10. Mandel, L.; Kaynar, A.; Wazen, J. Pneumoparotid: A case report. *Oral Surg. Oral Med. Oral Pathol.* **1991**, *72*, 22–24. [CrossRef]
11. Piette, E.; Walker, R.T. Pneumoparotid during dental treatment. *Oral Surg. Oral Med. Oral Pathol.* **1991**, *72*, 415–417. [CrossRef]
12. Takenoshita, Y.; Kawano, Y.; Oka, M. Pneumoparotis, an unusual occurrence of parotid gland swelling during dental treatment. Report of a case with a review of the literature. *J. Cranio-Maxillofacial Surg.* **1991**, *19*, 362–365. [CrossRef]
13. Krief, O.; Gomori, J.M.; Gay, I. CT of pneumoparotitis. *Comput. Med Imaging Graph.* **1992**, *16*, 39–41. [CrossRef]
14. Curtin, J.J.; Ridley, N.T.; Cumberworth, V.L.; Glover, G.W. Pneumoparotitis. *J. Laryngol. Otol.* **1992**, *106*, 178–179. [CrossRef]
15. Ferlito, A.; Andretta, M.; Baldan, M.; Candiani, F. Non-occupational recurrent bilateral pneumoparotitis in an adolescent. *J. Laryngol. Otol.* **1992**, *106*, 558–560. [CrossRef]
16. Brown, F.H.; Ogletree, R.C.; Houston, G.D. Pneumoparotitis Associated With the Use of an Air-Powder Prophylaxis Unit. *J. Periodontol.* **1992**, *63*, 642–644. [CrossRef]
17. Birzgalis, A.R.; Curley, J.W.; Camphor, I. Pneumoparotitis, subcutaneous emphysema and pleomorphic adenoma. *J. Laryngol. Otol.* **1993**, *107*, 349–351. [CrossRef]
18. McDuffie, M.W.; Brown, F.H.; Raines, W.H. Pneumoparotitis with orthodontic treatment. *Am. J. Orthod. Dentofac. Orthop.* **1993**, *103*, 377–379. [CrossRef]
19. Cook, J.N.; Layton, S.A. Bilateral parotid swelling associated with chronic obstructive pulmonary disease. A case of pneumoparotid. *Oral Surgery Oral Med. Oral Pathol.* **1993**, *76*, 157–158. [CrossRef]
20. Nassimbeni, G.; Ventura, A.; Boehm, P.; Guastalla, P.; Zocconi, E. Self-Induced Pneumoparotitis. *Clin. Pediatr.* **1995**, *34*, 160–162. [CrossRef]
21. Goguen, L.A.; April, M.M.; Karmody, C.S.; Carter, B.L. Self-induced Pneumoparotitis. *Arch. Otolaryngol. Head Neck Surg.* **1995**, *121*, 1426–1429. [CrossRef] [PubMed]
22. Ros, S.P.; Tamayo, R.C. A case of swollen parotid gland. *Pediatr. Emerg. Care* **1996**, *12*, 205–206. [CrossRef] [PubMed]
23. Gudlaugsson, Ó.; Geirsson, Á.J.; Benediktsdóttir, K. Pneumoparotitis: A new diagnostic technique and a case report. *Ann. Otol. Rhinol. Laryngol.* **1998**, *107*, 356–358. [CrossRef]
24. Alcalde, R.E.; Ueyama, Y.; Lim, D.J.; Matsumura, T. Pneumoparotid: Report of a case. *J. Oral Maxillofac. Surg.* **1998**, *56*, 676–680. [CrossRef]
25. Golz, A.; Joachims, H.Z.; Netzer, A.; Westerman, S.T.; Gilbert, L.M. Pneumoparotitis: Diagnosis by computed tomography. *Am. J. Otolaryngol.* **1999**, *20*, 68–71. [CrossRef]
26. Sittel, C.; Jungehülsing, M.; Fischbach, R. High-resolution magnetic resonance imaging of recurrent pneumoparotitis. *Ann. Otol. Rhinol. Laryngol.* **1999**, *108*, 816–818. [CrossRef]
27. Kirsch, C.M.; Shinn, J.; Porzio, R.; Trefelner, E.; Kagawa, F.T.; Wehner, J.H.; Jensen, W.A. Pneumoparotid due to spirometry. *Chest* **1999**, *116*, 1475–1478. [CrossRef]
28. Martín-Granizo, R.; Herrera, M.; García-González, D.; Mas, A. Pneumoparotid in childhood: Report of two cases. *J. Oral Maxillofac. Surg.* **1999**, *57*, 1468–1471. [CrossRef]
29. Han, S.; Isaacson, G. Recurrent Pneumoparotid: Cause and Treatment. *Otolaryngol. Neck Surg.* **2004**, *131*, 758–761. [CrossRef]
30. Apaydin, M.; Sarsilmaz, A.; Calli, C.; Erdogan, N.; Varer, M.; Uluç, E. Giant pneumoparotitis. *Eur. J. Radiol. Extra* **2004**, *52*, 17–20. [CrossRef]
31. Grainger, J.; Saravanappa, N.; Courteney-Harris, R.G. Bilateral pneumoparotid. *Otolaryngol. Head Neck Surg.* **2006**, *134*, 531–532. [CrossRef]
32. Prabhu, S.P.; Tran, B. Pneumoparotitis. *Pediatr. Radiol.* **2008**, *38*, 1144. [CrossRef]
33. Luaces, R.; Ferreras, J.; Patiño, B.; García-Rozado, Á.; Vazquez, I.; López-Cedrún, J.L. Pneumoparotid: A Case Report and Review of the Literature. *J. Oral Maxillofac. Surg.* **2008**, *66*, 362–365. [CrossRef]
34. Faure, F.; Gaudon, I.P.; Tavernier, L.; Khalfallah, S.A.; Folia, M. A rare presentation of recurrent parotid swelling: Self-induced parotitis. *Int. J. Pediatr. Otorhinolaryngol. Extra* **2009**, *4*, 29–31. [CrossRef]
35. Kyung, S.K.; Heurtebise, F.; Godon, A.; Rivière, M.-F.; Coatrieux, A. Head-neck and mediastinal emphysema caused by playing a wind instrument. *Eur. Ann. Otorhinolaryngol. Head Neck Dis.* **2010**, *127*, 221–223. [CrossRef]

36. Mukundan, D.; Jenkins, O. A Tuba Player with Air in the Parotid Gland. *N. Engl. J. Med.* **2009**, *360*, 710. [CrossRef]
37. van Ardenne, N.; Kurotova, A.; Boudewyns, A. Pneumoparotid: A rare cause of parotid swelling in a 7-year-old child. *B-ENT* **2011**, *7*, 297–300.
38. Ghanem, M.; Brown, J.; McGurk, M. Pneumoparotitis: A diagnostic challenge. *Int. J. Oral Maxillofac. Surg.* **2012**, *41*, 774–776. [CrossRef]
39. Potet, J.; Arnaud, F.-X.; Valbousquet, L.; Ukkola-Pons, E.; Donat-Weber, G.; Thome, A.; Péroux, E.; Teriitehau, C.; Baccialone, J. Pneumoparotid, a rare diagnosis to consider when faced with unexplained parotid swelling. *Diagn. Interv. Imaging* **2013**, *94*, 95–97. [CrossRef]
40. Konstantinidis, I.; Chatziavramidis, A.; Constantinidis, J. Conservative management of bilateral pneumoparotitis with sialendoscopy and steroid irrigation. *BMJ Case Rep.* **2014**, *2014*. [CrossRef]
41. Abdullayev, R.; Saral, F.C.; Kucukebe, O.B.; Sayiner, H.S.; Bayraktar, C.; Akgun, S. Bilateral parotitis in a patient under continuous positive airway pressure treatment. *Braz. J. Anesth. Engl. Ed.* **2016**, *66*, 661–663. [CrossRef]
42. Cabello, M.; Macias, E.; Fernández-Flórez, A.; Martínez-Martínez, M.; Cobo, J.; de Carlos, F. Pneumoparotid associated with a mandibular advancement device for obstructive sleep apnea. *Sleep Med.* **2015**, *16*, 1011–1013. [CrossRef]
43. Alnæs, M.; Furevik, L.L. Pneumoparotitis. *Tidsskr. Nor. Laegeforen.* **2017**, *137*, 544. [CrossRef]
44. Goates, A.J.; Lee, D.J.; Maley, J.; Lee, P.C.; Hoffman, H.T. Pneumoparotitis as a complication of long-term oronasal positive airway pressure for sleep apnea. *Head Neck* **2017**, *40*, E5–E8. [CrossRef]
45. Lagunas, J.G.; Fuertes, A.F. Self-induced parapharyngeal and parotid emphysema: A case of pneumoparotitis. *Oral Maxillofac. Surg. Cases* **2017**, *3*, 81–85. [CrossRef]
46. Yamazaki, H.; Kojima, R.; Nakanishi, Y.; Kaneko, A. A Case of Early Pneumoparotid Presenting With Oral Noises. *J. Oral Maxillofac. Surg.* **2018**, *76*, 67–69. [CrossRef]
47. Lee, K.P.; James, V.; Ong, Y.-K.G. Emergency Department Diagnosis of Idiopathic Pneumoparotitis with Cervicofacial Subcutaneous Emphysema in a Pediatric Patient. *Clin. Pract. Cases Emerg. Med.* **2017**, *1*, 399–402. [CrossRef]
48. House, L.K.; Lewis, A.F. Pneumoparotitis. *Clin. Exp. Emerg. Med.* **2018**, *5*, 282–285. [CrossRef]
49. Ambrosino, R.; Lan, R.; Romanet, I.; Le Roux, M.-K.; Gallucci, A.; Graillon, N.; Audrey, G.; Nicolas, G. Severe idiopathic pneumoparotitis: Case report and study review. *Int. J. Pediatr. Otorhinolaryngol.* **2019**, *125*, 196–198. [CrossRef]
50. Flores-Orozco, E.I.; Tiznado-Orozco, G.E.; Díaz-Peña, R.; Orozco, E.I.F.; Galletti, C.; Gazia, F.; Galletti, F. Effect of a Mandibular Advancement Device on the Upper Airway in a Patient With Obstructive Sleep Apnea. *J. Craniofacial Surg.* **2020**, *31*, e32–e35. [CrossRef]
51. Cammaroto, G.; Galletti, C.; Galletti, F.; Galletti, B.; Galletti, C.; Gay-Escoda, C. Mandibular advancement devices vs nasal-continuous positive airway pressure in the treatment of obstructive sleep apnoea. Systematic review and meta-analysis. *Med. Oral Patol. Oral Y Cir. Bucal* **2017**, *22*, e417–e424. [CrossRef]
52. Giudice, A.L.; Galletti, C.; Gay-Escoda, C.; Leonardi, R. CBCT assessment of radicular volume loss after rapid maxillary expansion: A systematic review. *J. Clin. Exp. Dent.* **2018**, *10*, e484–e494. [CrossRef]
53. Lasboo, A.A.; Nemeth, A.; Russell, E.J.; Siegel, G.J.; Karagianis, A. The use of the "puffed-cheek" computed tomography technique to confirm the diagnosis of pneumoparotitis. *Laryngoscope* **2010**, *120*, 967–969. [CrossRef]
54. Meerleer, K.D.; Hermans, R. Images in clinical radiology: Pneumoparotitis. *Jbr-Btr* **2005**, *88*, 248.
55. Galletti, B.; Freni, F.; Gazia, F.; Nicastro, V.; Abita, P.; Sireci, F.; Galletti, F. A rare case of a huge pleomorphic adenoma of minor salivary glands in the parapharyngeal space. *Euromediterranean Biomed. J.* **2019**, *14*, 7–10.
56. Freni, F.; Gazia, F.; Stagno d'Alcontres, F.; Galletti, B.; Galletti, F. Use of botulinum toxin in Frey's syndrome. *Clin. Case Rep.* **2019**, *31*, 482–485. [CrossRef]
57. Moltrecht, M.; Michel, O. The woman behind Frey's syndrome: The tragic life of Lucja Frey. *Laryngoscope* **2004**, *114*, 2205–2209. [CrossRef]

MDPI
St. Alban-Anlage 66
4052 Basel
Switzerland
Tel. +41 61 683 77 34
Fax +41 61 302 89 18
www.mdpi.com

International Journal of Environmental Research and Public Health Editorial Office
E-mail: ijerph@mdpi.com
www.mdpi.com/journal/ijerph

www.ingramcontent.com/pod-product-compliance
Lightning Source LLC
LaVergne TN
LVHW070701170726
843515LV00004B/455
9783036515045